AF380607

The M14 Rifle

**7.62MM
M14 and M14E2
M14SA**

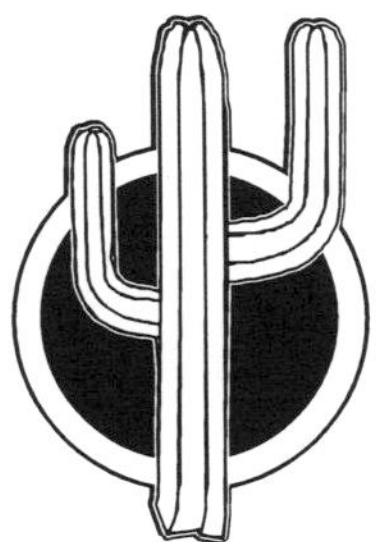

Desert Publications
El Dorado, AR 71731-1751 U. S. A.

The M14 Rifle
7.62MM, M14, M14E2 & M14SA

© 1978 by Desert Publications
215 S. Washington
El Dorado, AR 71731-1751
870-862-2077

ISBN 0-87947-015-1
15 14 13 12 11 10
Printed in U. S. A.

**Desert Publications is a division of
The DELTA GROUP, Ltd.
Direct all inquiries & orders to the above address.**

Neither the author nor the publisher assumes any responsibility for the use or misuse of the information contained in this book. This material was compiled for educational and entertainment purposes and one should not construe that any other purpose is suggested.

FIELD MANUAL

No. 23-8

HEADQUARTERS
DEPARTMENT OF THE ARMY
WASHINGTON, D.C., *7 May 1965*

A reprint of the original

U.S. RIFLE

7.62MM, M14 AND M14E2

CHAPTER 1

INTRODUCTION

1. Purpose and Scope

a. This manual is a guide for commanders and instructors in presenting instruction in the mechanical operation of the M14 and M14E2 rifles. It includes a detailed description of the rifle and its general characteristics; procedures for detailed disassembly and assembly; an explanation of functioning; a discussion of the types of stoppages and the immediate action applied to reduce them; a description of the ammunition; and instructions on the care, cleaning, and handling of each weapon and its ammunition.

b. Marksmanship training is covered in FM 23–71 and FM 23–16.

c. The material contained herein is applicable without modification to both nuclear and nonnuclear warfare.

d. Users of this manual are encouraged to submit recommended changes or comments to improve the publication. Comments should be keyed to the specific page, paragraph, and line of the text

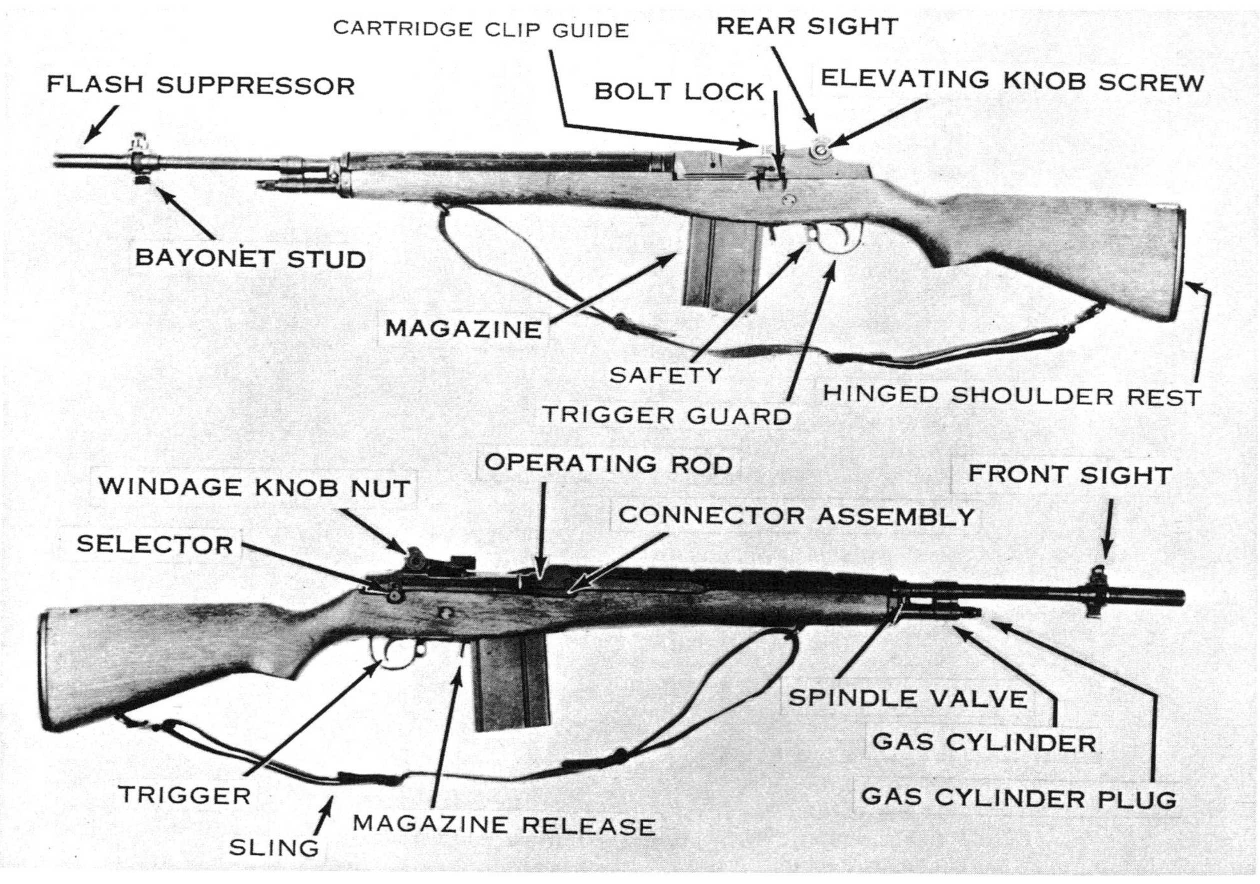

Figure 1. The M14 rifle.

in which the change is recommended. Reasons should be provided for each comment to insure understanding and complete evaluation. Comments should be forwarded direct to the Commandant, United States Army Infantry School, Fort Benning, Ga., 31905.

2. Importance of Mechanical Training

The rifle is the Infantryman's basic weapon. It gives him an individual and powerful capability for combat. To benefit the most from this capability, the Infantryman must develop two skills to an equal degree: he must be able to fire his weapon well enough to get hits on battlefield targets, and he must know enough about its working parts to keep it operating. The Infantryman attains his firing skill in marksmanship training. He learns how to keep his rifle in operable condition through mechanical training.

3. Description of the Rifles

a. M14 Rifle.
   (1) The U.S. rifle, 7.62mm, M14 (fig. 1) is a light-weight, air-cooled, gas-operated, magazine-fed, shoulder weapon. It is designed primarily for semiautomatic fire.
   (2) When employed as an automatic rifle, the selector and bipod M2 must be installed (fig. 2).
   (3) The flash suppressor is designed with a wide rib on the bottom to reduce muzzle climb and the amount of dust raised by muzzle blast.
   (4) The lug on the rear of the flash suppressor is used to secure a bayonet, a grenade launcher, and a blank firing attachment.
   (5) The spindle valve is used when launching a grenade to prevent gas operation of the rifle, thus avoiding damage to the weapon.

b. M14E2 Rifle.
   (1) The U.S. rifle, 7.62mm, M14E2 (fig. 3) is an air-cooled, gas-operated, magazine-fed, shoulder weapon. It is capable of semiautomatic or automatic fire; however, it is designed primarily for automatic fire. It features a stabilizer assembly, modified bipod, front and rear handgrip, straight line stock, and a rubber recoil pad.
   (2) The M14E2 stock group is the "straight line" type with a fixed rear handgrip and a folding front handgrip which lies flat along the bottom of the stock when not in use. The location of the front handgrip can be adjusted to one of five positions in 1-inch increments to accommodate all gunners. The rubber recoil pad reduces the effects of recoil. The hinged shoulder rest provides vertical control of the butt end of the rifle. The butt swivel pivots 90° to the left for ease of carrying.
   (3) The stabilizer assembly consists of a perforated steel sleeve which slides over the flash suppressor and is fastened to the muzzle over the bayonet lug by a screw and a locknut. The stabilizer provides muzzle stability and reduces recoil.

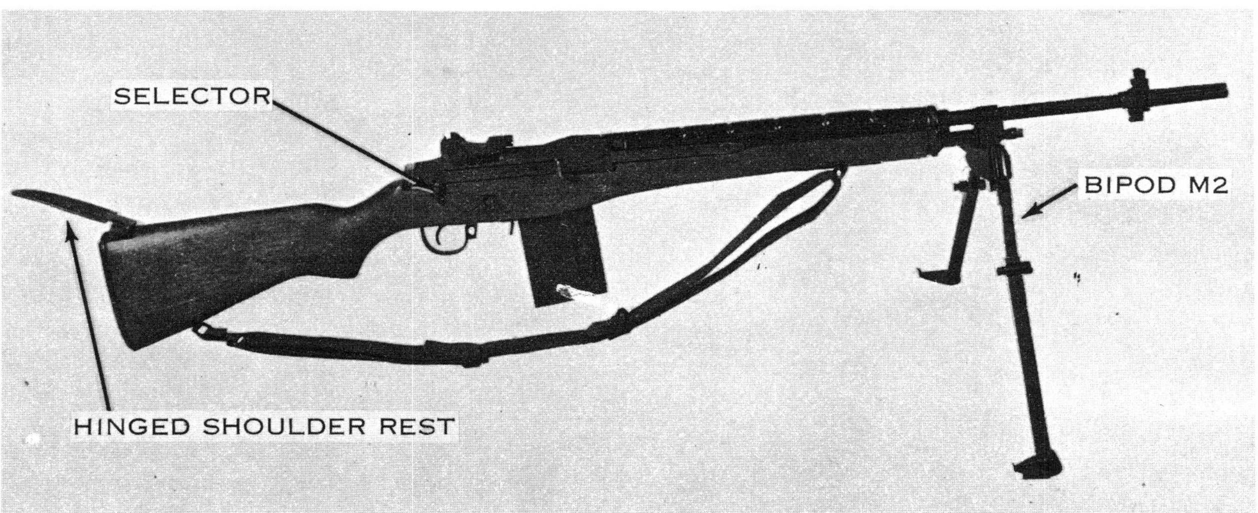

Figure 2. The M14 rifle with selector and M2 bipod.

(4) The M2 bipod is modified by the addition of a sling swivel and a longer pivot pin to accommodate the swivel.

(5) The M14E2 utilizes a sling with an extra hook assembly. The portion of the sling between the handgrip and the bipod provides additional muzzle control during firing. The portion of the sling between the front handgrip and the bipod allows the average firer, by applying rearward pressure on the front handgrip, to increase the pressure of the bipod on the ground to approximately 35 pounds, reducing dispersion considerably. When the weapon is carried at sling arms, the sling must be disconnected from the handgrip assembly.

4. General Data

Weights in Pounds (approximate):

M14 rifle with full magazine and cleaning equipment	11¼
M14 rifle with full magazine, cleaning equipment, selector, and bipod	13
Empty magazine	½
Full magazine (with ball ammunition)	1½
Cleaning equipment	⅔
M2 bipod	1¾
M14E2 rifle with full magazine	14½

Lengths in Inches (approx.):

M14, overall, with flash suppressor	44⅛
M14E2, overall, with stabilizer assembly	44⅛

Sights:

Front	fixed
Rear	adjustable (one click of elevation or windage moves the strike of the bullet .7 centimeter at 25 meters).

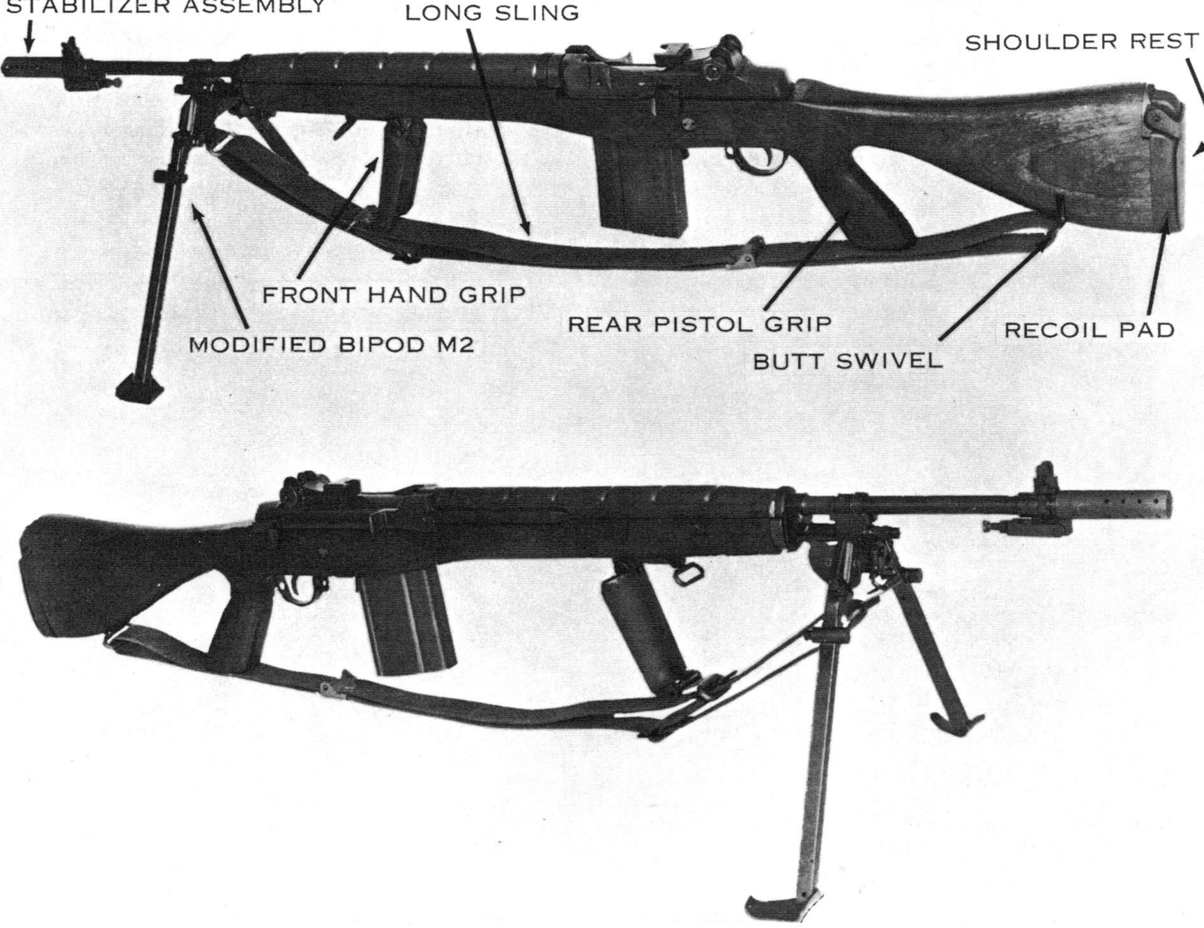

Figure 3. The M14E2 rifle (top—left side view; bottom—right side view).

Trigger Pull in Pounds:
  Minimum______________________ 5.5
  Maximum______________________ 7.5
*Muzzle Velocity*________________ 2,800 f.p.s. (853 m.p.s.).
Cyclic Rate of Fire (rounds per minute). 700–750
Rates of Fire. (These can be maintained without danger to the firer, or damage to the weapon):
  Semiautomatic (rounds per minute):
    1 minute______________ 40
    2 minutes_____________ 40
    5 minutes_____________ 30
    10 minutes____________ 20
    15 minutes____________ 20
    20 minutes____________ 20
    30 minutes (or more)____ 15
  Automatic (rounds per minute):
    1 minute______________ 60
    2 minutes_____________ 50
    5 minutes_____________ 40
    10 minutes____________ 30
    15 minutes____________ 30
    20 minutes____________ 25
    30 minutes (or more)____ 20
Range in Meters:
  Maximum effective (semiautomatic, without bipod). 460
  Maximum effective (semiautomatic, with bipod). *700
  Maximum effective (automatic, with bipod). **460
  Maximum______________________ 3725
*Ammunition*____________________ see chapter 6.
Definitions:
  Cyclic rate__________________ the rate at which the weapon fires automatically.
  Maximum effective range_____ the greatest distance at which a weapon may be expected to fire accurately to inflict casualties or damage.

*The bipod adds much stability to the rifle and enables the automatic rifleman to effectively engage targets semiautomatically in excess of 460 meters.

**Enemy squad formations and hasty crew-served weapons emplacements may be effectively engaged up to this range; bunker apertures, windows and like targets, which require precise accuracy, can best be engaged using semiautomatic fire.

CHAPTER 2

MECHANICAL TRAINING

5. General

a. The individual soldier is authorized to disassemble his rifle to the extent called field stripping. Chart I shows the parts he is permitted to disassemble with and without supervision. The amount of disassembly he is permitted to perform without supervision is adequate for normal maintenance.

b. The frequency of disassembly and assembly should be kept to a minimum consistent with maintenance and instructional requirements. Constant disassembly causes excessive wear of the parts and leads to their early unserviceability and to inaccuracy of the weapon.

c. The rifle has been designed to be taken apart and put together easily. No force is needed if it is disassembled and assembled correctly. The parts of one rifle, *except the bolt,* may be interchanged with those of another *when necessary. Bolts should never be interchanged for safety reasons.*

d. As the rifle is disassembled, the parts should be laid out from left to right, on a clean surface and in the order of removal. This makes assembly easier because the parts are assembled in the reverse order of disassembly. The names of the parts (nomenclature) should be taught along with disassembly and assembly to make further instruction on the rifle easier to understand.

6. Clearing the Rifle

The first step in handling any weapon is to clear it. To clear the rifle, first attempt to engage the safety. (If unable to place the safety in the safe position, continue with the second step of removing the magazine.) Remove the magazine by placing the right thumb on the magazine latch and curl the remaining fingers around the front of the magazine. Press in on the magazine latch, rotate the base of the magazine toward the muzzle end of the rifle (fig. 4), and remove it from the magazine well. With the knife edge of the right hand, pull the operating rod handle all the way to the rear, reach across the receiver with the right thumb and press in on the bolt lock (fig. 5). Verify the

Chart I. Disassembly Authorization

Part	Individual soldier	Armorer	Maintenance personnel
SEPARATION INTO THREE MAIN GROUPS	X		
DISASSEMBLY:			
BARREL AND RECEIVER GROUP	X		
Front sight			X
Rear sight		X	
Flash suppressor			X
Spindle valve			X
Sear release		X	
Selector and selector shaft lock		X	
Bipod M2	X		
Connector assembly (spring and plunger)			X
Bolt lock		X	
Cartridge clip guide			X
Operating rod guide			X
Barrel from receiver			X
Stabilizer assembly M14E2	X		
STOCK GROUP:			
Stock liner			X
Upper sling swivel bracket			X
Stock ferrule			X
MAGAZINE	X		
BOLT		X	
Bolt roller from bolt stud			X
FIRING MECHANISM		X	
Magazine latch			X
Sear from trigger			X

safety, tilt the rifle, and look inside the chamber and receiver to insure that they contain no rounds.

7. Disassembly Into Three Main Groups

a. The three main groups are the firing mechanism, the barrel and receiver, and the stock.

b. After the rifle is cleared, the operating parts should be forward for disassembly. To do this, pull back on the operating rod handle and allow the bolt to go forward.

c. To remove the firing mechanism, grasp the rear of the trigger guard with the thumb and forefinger of your right hand and pull downward and outward until the mechanism is released (fig. 6). Lift out the firing mechanism.

d. To separate the barrel and receiver from the stock, lay the weapon on a flat surface with the sights up and muzzle to the left. Grasp the receiver with the left hand over the rear sight and raise the rifle a few inches. With the right hand, strike down on and grasp the small of the stock, separating the barrel and receiver from the stock. The three main groups are shown in figure 7.

e. The components of the M14E2 rifle are shown in figure 8.

8. Assembly of the Three Main Groups

a. Place the barrel and receiver group on a flat surface, sights down. Pick up the stock group and engage the stock ferrule in the front band, then lower the stock group onto the barrel and receiver group.

b. Open the trigger guard and place the firing mechanism straight down into the receiver, making sure that the guide rib on the firing mechanism enters the recess in the receiver (fig. 9). Place the butt of the weapon on the left thigh, sights to the left, insuring the trigger guard has cleared the trigger. With the palm of the right hand, strike the trigger guard fully engaging it to the receiver.

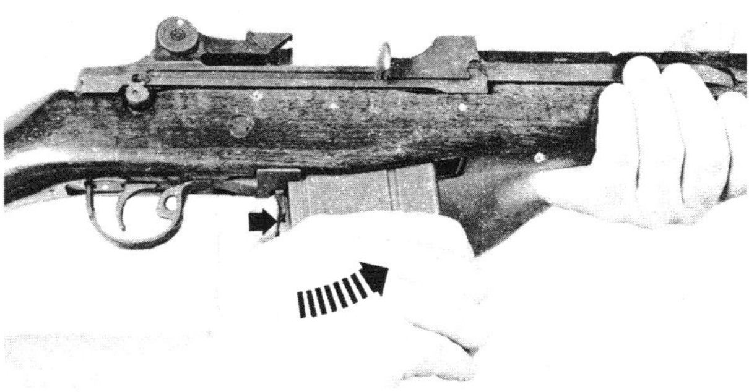

Figure 4. Removing the magazine.

Figure 5. Locking the bolt to the rear.

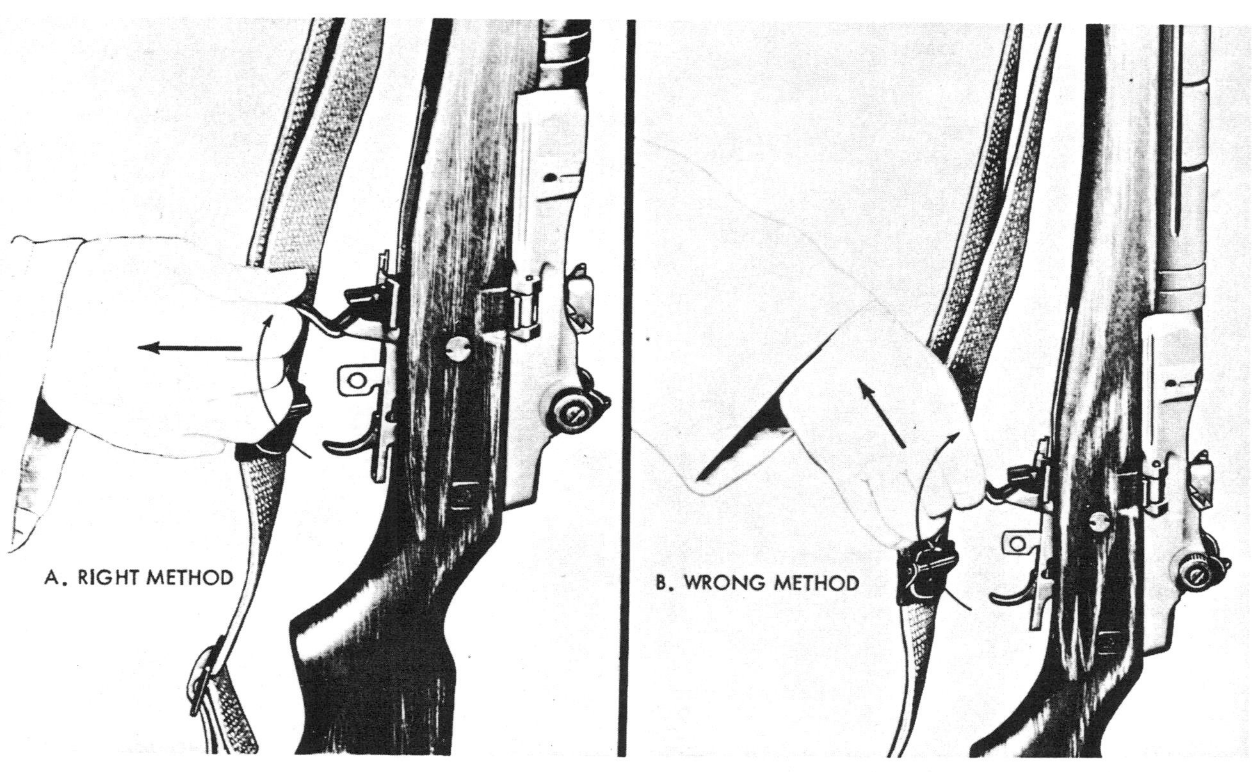

Figure 6. Removing firing mechanism.

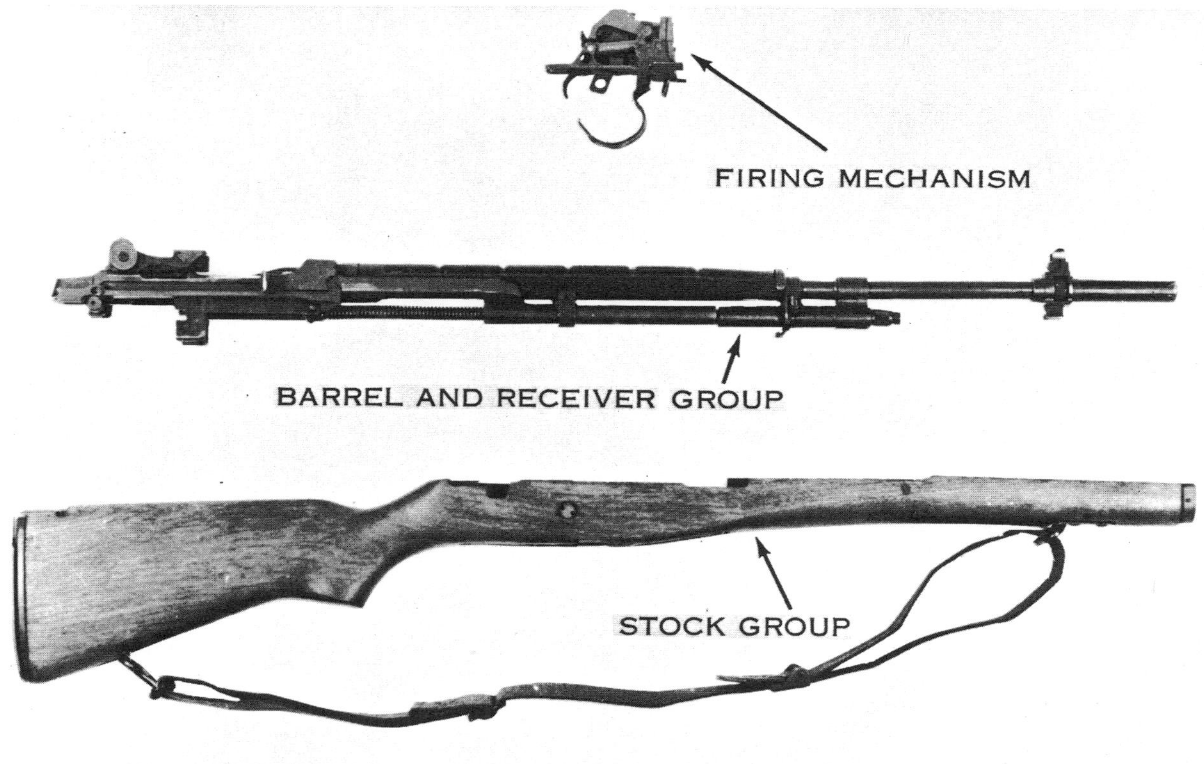

Figure 7. The three main groups.

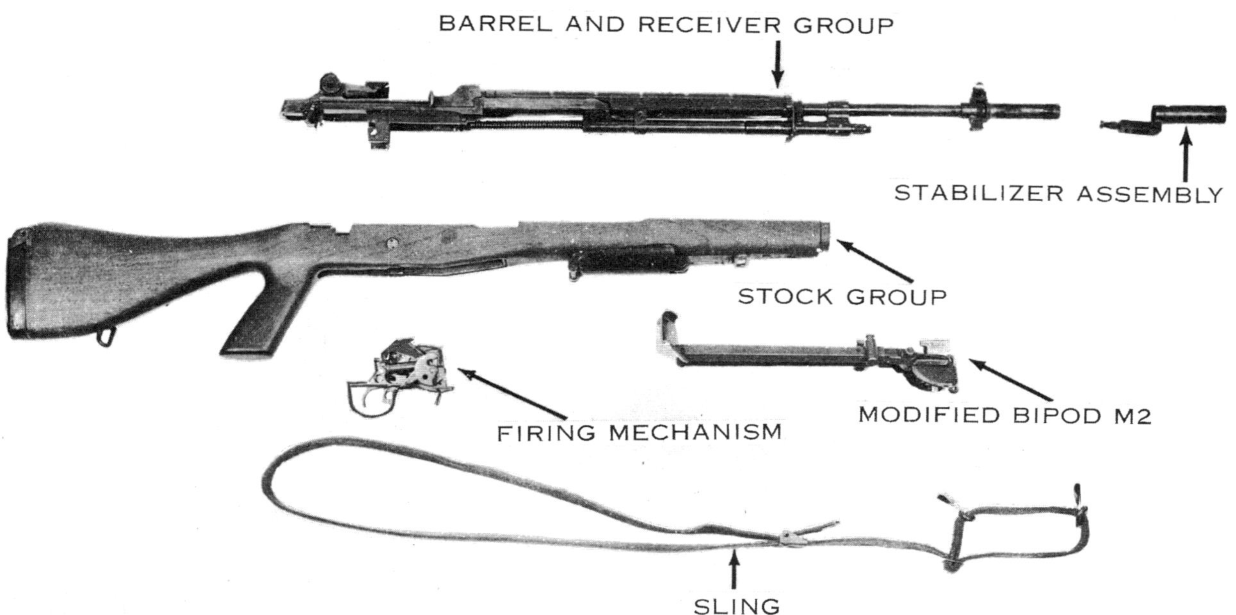

Figure 8. Components of the M14E2 rifle.

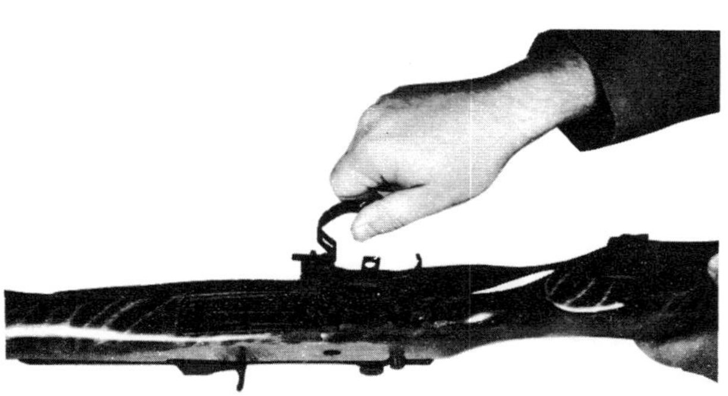

Figure 9. Replacing the firing mechanism.

9. Disassembly of the Barrel and Receiver Group

a. Removing the Connector Assembly. Place the barrel and receiver group on its left side with the operating rod handle up and the muzzle away from you. On rifles modified for selective firing, press in and turn the selector until the face marked "A" is toward the windage knob (fig. 10). With the bolt closed, place the right thumb on the rear of the connector assembly, the first finger on the sear release bracket and the second finger inside the rear of the receiver (fig. 11). Push forward with the thumb until the forward end of the assembly can be lifted off the connector lock with the thumb and forefinger of the left hand (2, fig. 11). (Note that the rifle shown in 1, 2, and 3, fig. 11 has not been modified for selective firing.) Turn the con-nector assembly (3, fig. 11) clockwise until the elongated hole in the connector assembly is alined with the elongated stud on the sear release. Lower the front end of the connector assembly and lift the rear end off the elongated stud of the sear release.

b. Removing the Operating Rod Spring and Operating Rod Spring Guide. Place the barrel and receiver group on a flat surface, sights down, muzzle to the left. With your left hand, pull toward the muzzle on the operating rod spring to relieve pressure on the connector lock (1, fig. 12). With your right forefinger, pull the connector lock toward you and, allowing the operating rod spring to expand slowly, disconnect and remove the operating rod spring and operating rod spring guide (2, fig. 12). Separate these two parts.

c. Removing the Operating Rod. Turn the barrel and receiver group so the sights are up and the muzzle is pointing away from you. Pull back the operating rod handle until the guide lug on its inside surface is alined with the disassembly notch on the right side of the receiver. Rotate the operating rod downward and outward, then pull it to the rear, disengaging it from the operating rod guide (fig. 13).

d. Removing the Bolt. Grasp the bolt by the roller and, while sliding it forward, lift it upward and outward to the right front with a slight rotating motion (fig. 14).

e. Rifle Field Stripped. The parts of the barrel and receiver group in their order of disassembly are shown in figure 15.

10. Assembly of the Barrel and Receiver Group

a. Replacing the Bolt. Place the barrel and receiver on the table, sights up, muzzle pointing away from you. Hold the bolt by the roller and locking lug and place the rear of the bolt on the bridge of the receiver, firing pin tang pointed down. Turn the bolt slightly counterclockwise until the tang of the firing pin clears the bridge. Guide the left locking lug of the bolt into its groove on the left side of the receiver. Lower the right locking lug on its bearing surface and slide the bolt halfway to the rear.

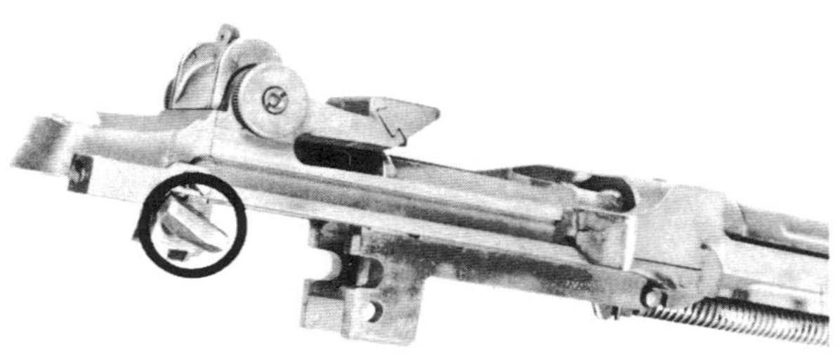

Figure 10. Position of the selector for removing the connector assembly (rifle modified for selective firing).

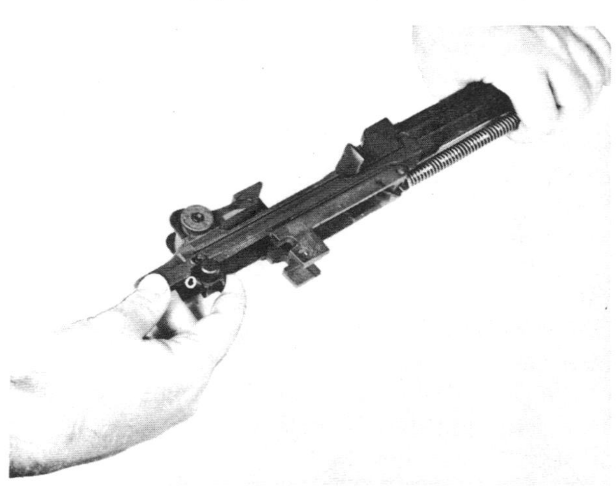

1

Figure 11. Removing the connector assembly.

2

Figure 11—Continued.

3

Figure 11—Continued.

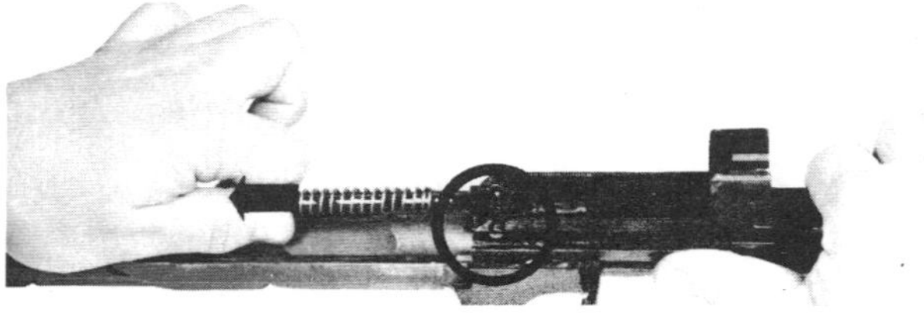

1

Figure 12. Removing the operating rod spring and operating rod spring guide.

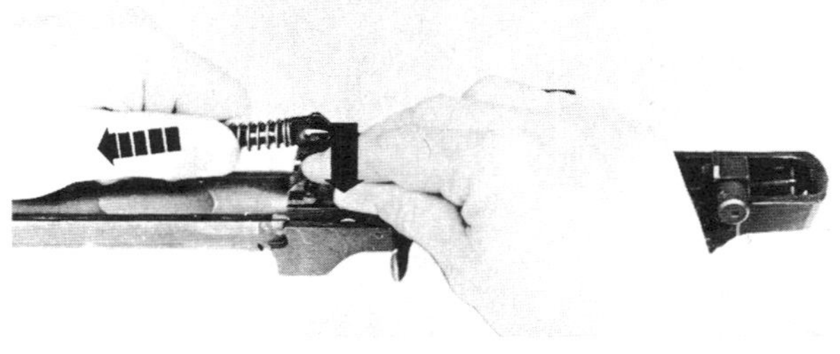

2

Figure 12—Continued.

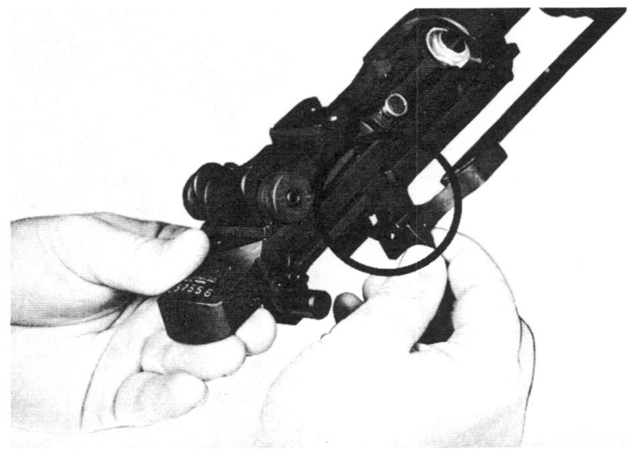

Figure 13.   Removing operating rod.

Figure 14.   Removing the bolt.

b. Replacing the Operating Rod. Holding the operating rod at the handle, place the front end into the operating rod guide, and position the rod so that the recess in the hump fits over the bolt roller. Turn the operating rod to the left until the guide lug fits into the disassembly notch on the receiver, then move the operating rod forward until the bolt is closed.

c. Replacing the Operating Rod Spring and Operating Rod Spring Guide. Turn the barrel and receiver over so the sights are down and the muzzle is to the left. Place the operating rod spring guide into the operating rod spring, hump up, and feed the loose end of the spring into the operating rod. Grasp the spring and guide with the left hand and compress the spring until the hole in the guide can be alined with the connector lock. Lower the guide and push the connector lock in with the right thumb (fig. 16).

d. Replacing the Connector Assembly. Place the barrel and receiver on its side with the operating rod handle up, muzzle away from you. Place the elongated hole in the rear of the connector assembly on the elongated stud on the sear release (1, fig. 17). Place the thumb of the right hand on the rear of the connector assembly, the first finger on the sear release bracket, and the second finger inside the rear of the receiver. Pushing toward the muzzle with the right thumb and with the thumb and first finger of the left hand, turn the front of the connector counterclockwise until it can be snapped onto the connector lock (2, fig. 17).

11

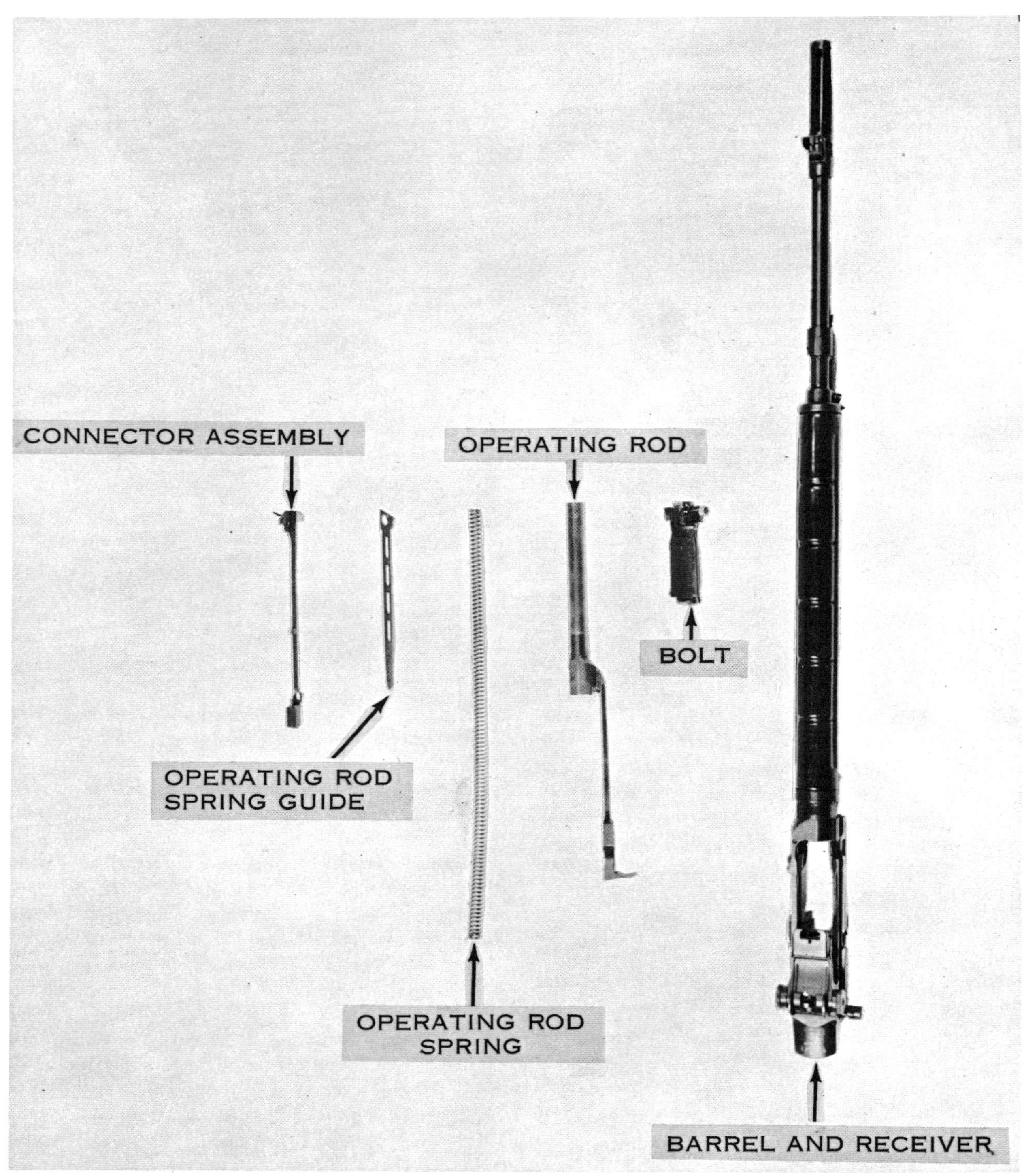

Figure 15. Parts of the barrel and receiver group in the order of disassembly.

11. Disassembly of the Gas System and Handguard

(fig. 18)

Note. Under normal usage the gas cylinder should not be disassembled as long as the gas piston slides freely within the cylinder when the barrel is tilted end-for-end from an upright position (bolt should be locked to the rear). Disassembly of the gas cylinder is sometimes necessary after the weapon has been subjected to extreme climatic conditions.

a. Gas System. Using the wrench of the combination tool, loosen and remove the gas cylinder plug. Tilt the muzzle down and remove the gas piston from the gas cylinder. Unscrew the gas cylinder lock and slide the lock and cylinder forward so that the gas port is exposed.

b. Handguard. Slip the front band forward toward the front sight. Push the handguard toward the front sight and lift it from the barrel.

12. Assembly of the Gas System and Handguard

a. Handguard. Place the rifle on a flat surface, sights up and muzzle to the right. Engage the ends of the band on the handguard with the front (muzzle) end of the slots that are on the rear of the barrel and *slide* the handguard rearward. (*Do not snap or force* the handguard into its installed position.) Replace the front band.

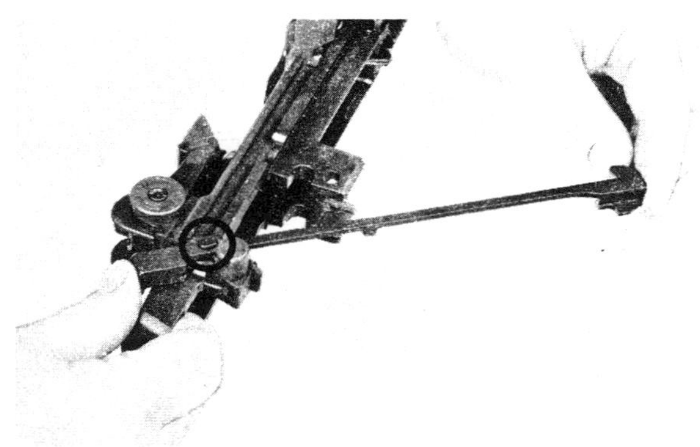

1

Figure 17. Replacing the connector assembly.

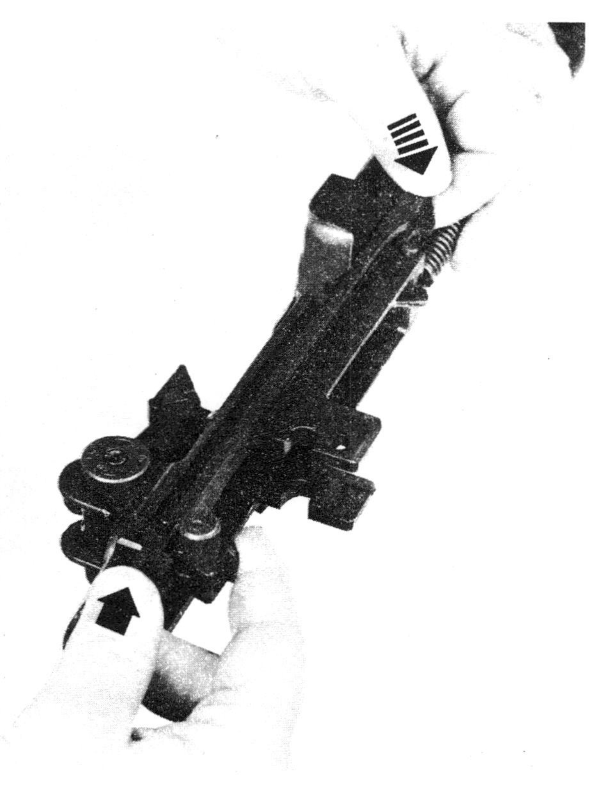

2

Figure 17—Continued.

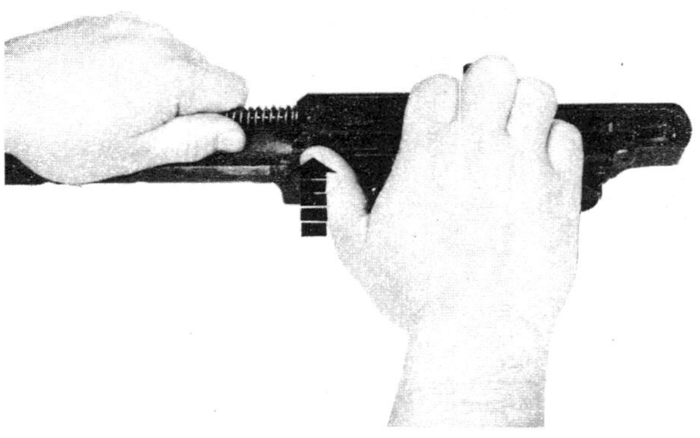

Figure 16. Replacing the operating rod spring and operating rod spring guide.

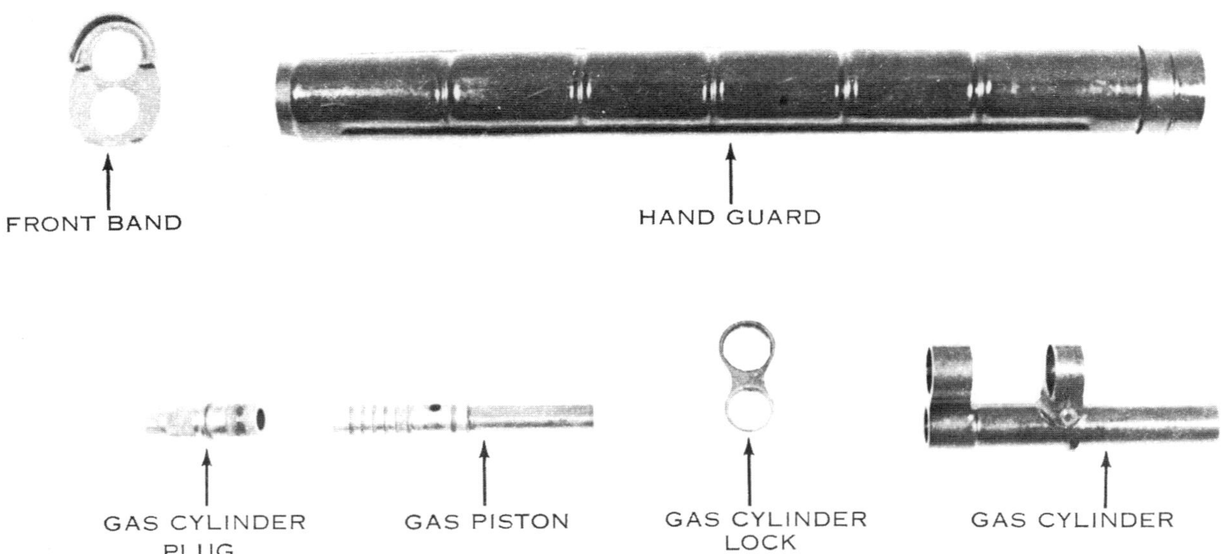

Figure 18. Parts of the gas system; handguard and front band.

b. Gas System. Slide the gas cylinder rearward through the front band. Tighten the gas cylinder lock by hand to its fully assembled position, then back it off until the loop is alined with the gas cylinder. Replace the gas piston with the flat part toward the barrel and the open end toward the muzzle. When the gas piston *is properly seated,* it will protrude one and one-half inches below the gas cylinder (fig. 19). Replace the gas cylinder plug and tighten it securely with the wrench of the combination tool.

13. Removing the Stabilizer Assembly of the M14E2 Rifle

To remove the stabilizer assembly, use the wrench end of the combination tool to loosen the locknut. Then slide the combination tool over the screw and loosen it. Swing the yoke away from the bayonet lug, and slide the stabilizer assembly off the flash suppressor (fig. 20).

14. Replacing the Stabilizer Assembly of the M14E2 Rifle

To replace the stabilizer assembly slide it over the flash suppressor, swing the yoke over the bayonet lug, and tighten the screw with the combination tool (fig. 21). Slide the combination tool over the head of the screw and place it over the locknut.

15. Disassembly of the Magazine

a. Use a pointed object to raise the rear of the magazine base (fig. 22) until the indentation on the base is clear of the magazine. Grasp the magazine with either hand, with one finger of the hand covering the base. Remove the base and guide the spring, one coil at a time, to clear the retaining lips of the magazine.

b. Remove and separate the magazine spring and follower. Figure 23 shows the parts of the magazine.

16. Assembly of the Magazine

Reposition the spring inside the follower with the rectangular-shaped end of the spring against the rear of the follower, and replace the follower and spring inside the magazine. Be sure to fully seat the follower. Replace the magazine base (fig. 24).

Note. The bolt, rear sight, and the firing mechanism *will not be* disassembled by the individual under any circumstances (chart I).

Figure 19. Gas piston properly seated.

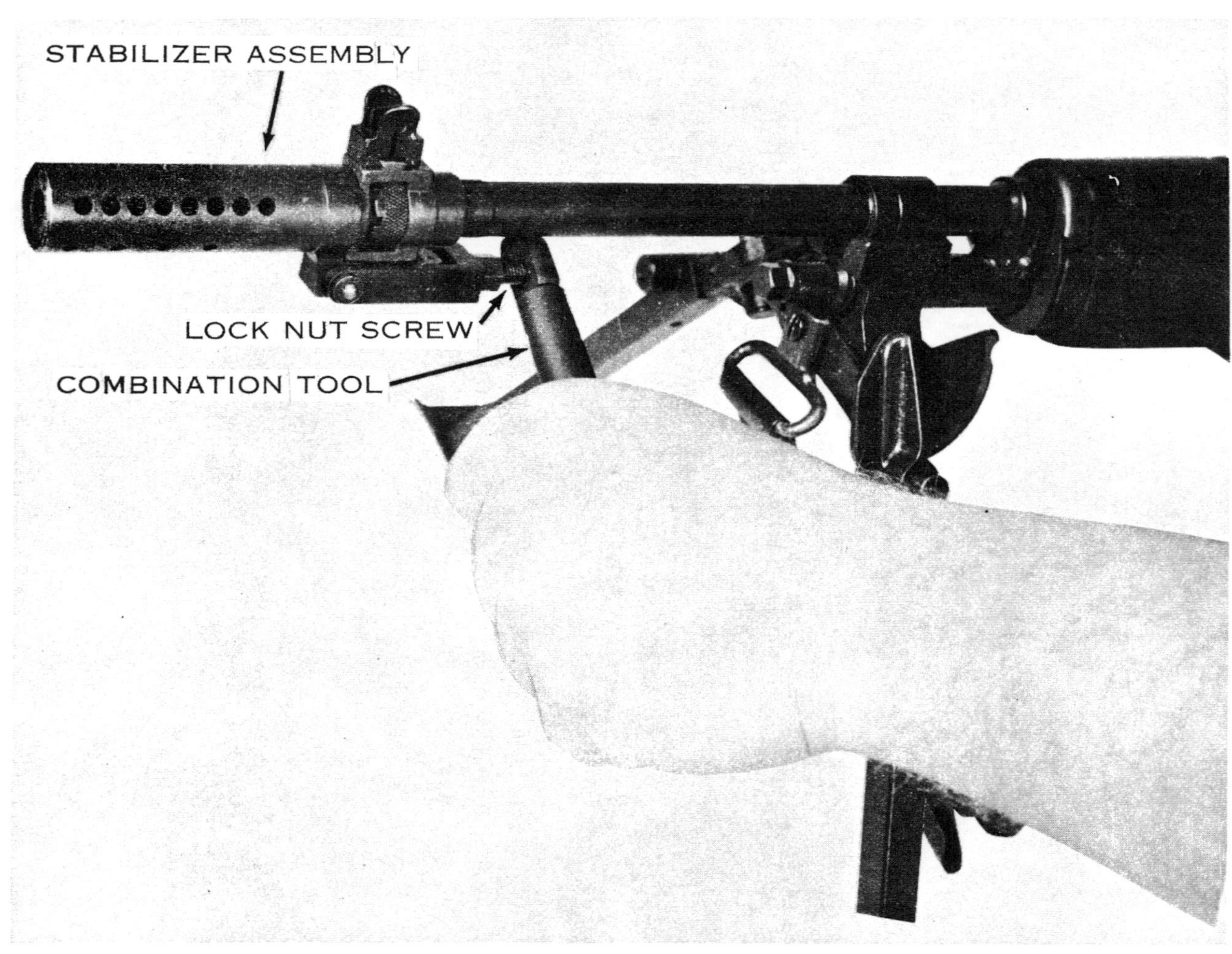

Figure 20. Removing the stabilizer assembly.

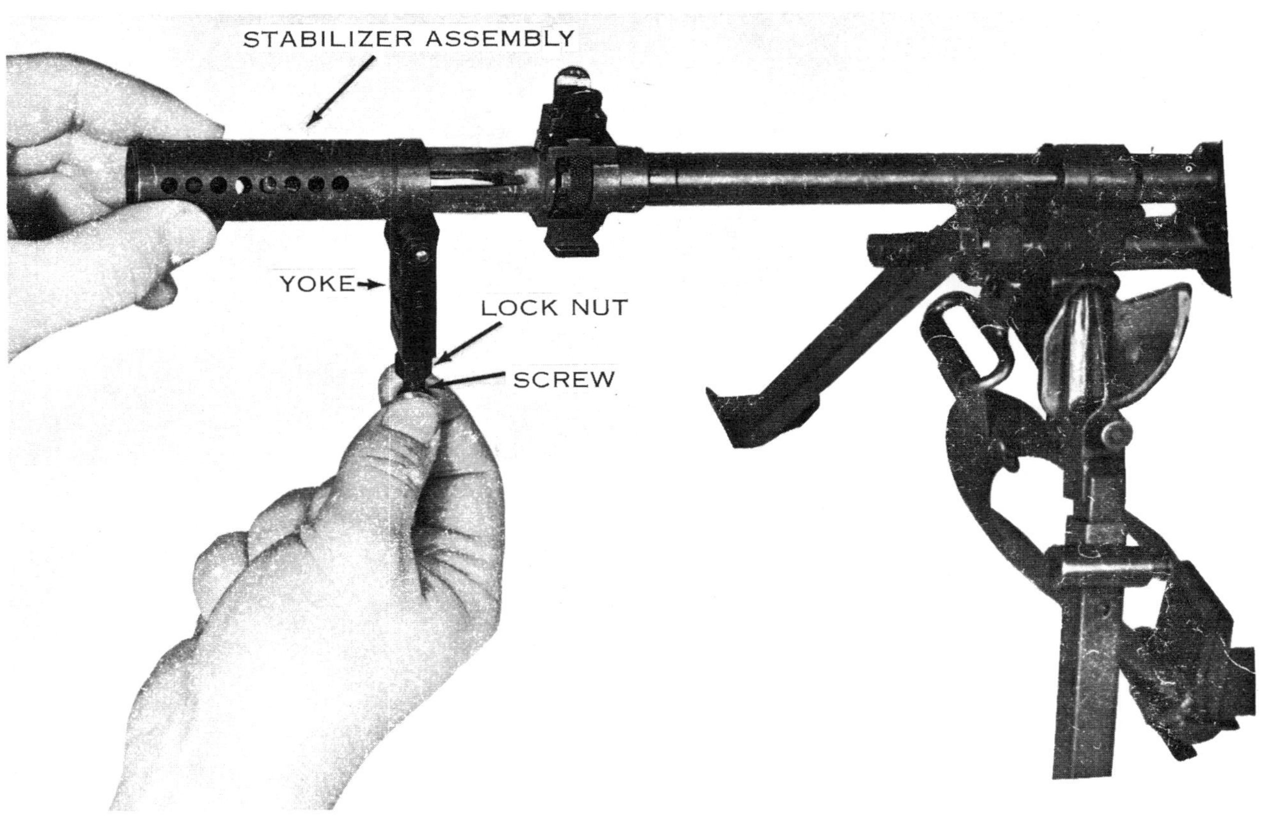

Figure 21. Replacing the stabilizer assembly.

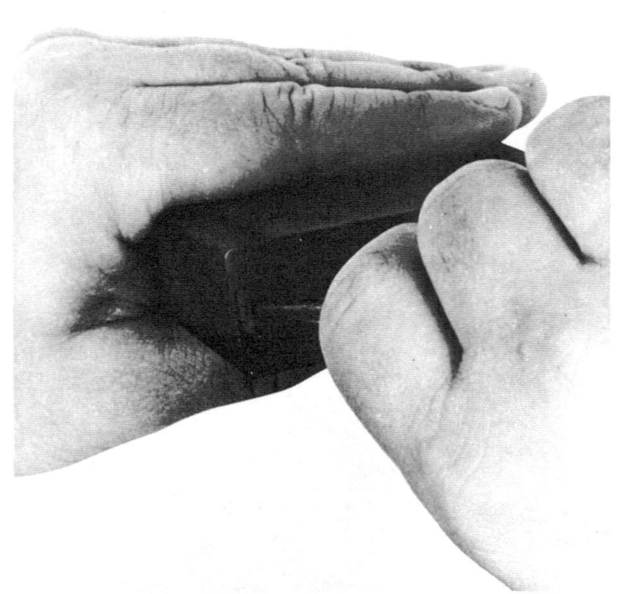

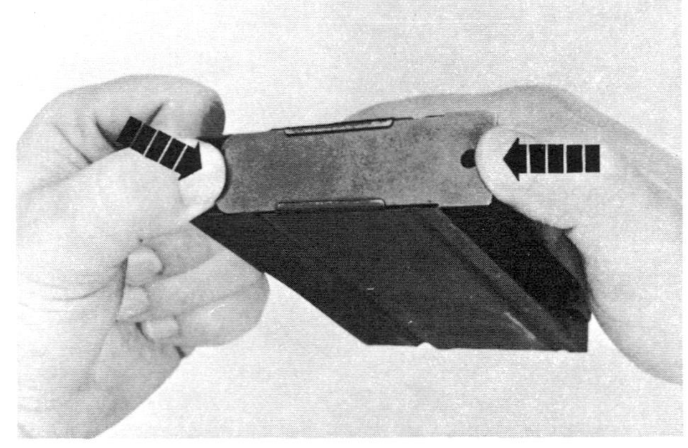

Figure 24. Replacing the magazine base.

Figure 22. Removing the base of the magazine.

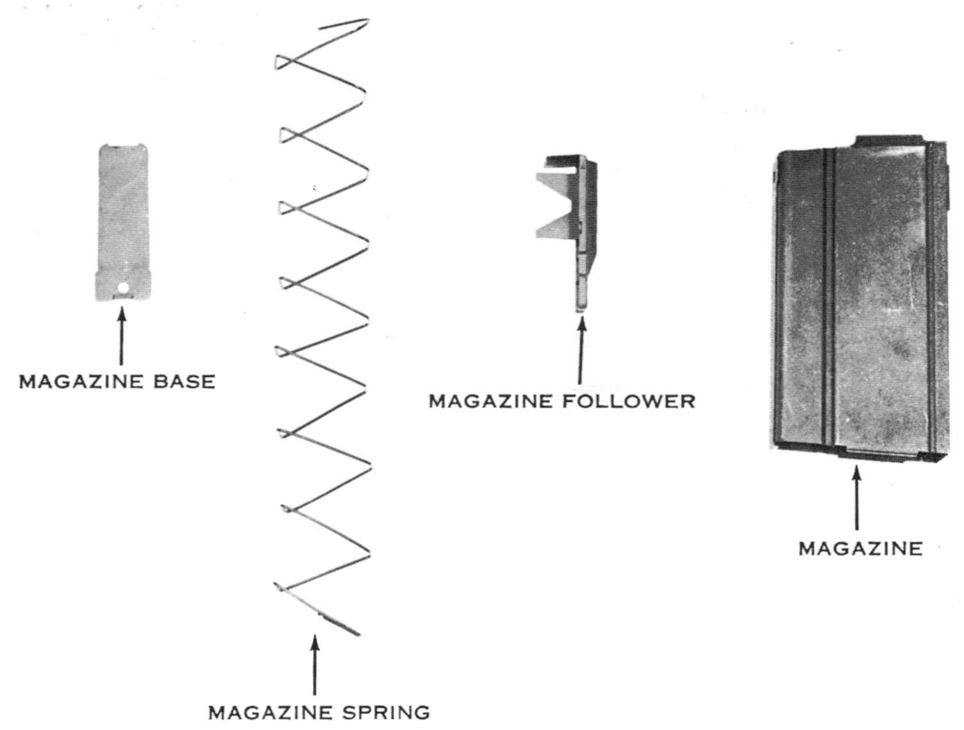

Figure 23. Parts of the magazine.

CHAPTER 3

OPERATION AND FUNCTIONING

Section I. OPERATION

17. Loading the Magazine (Out of the Rifle)

a. Place each round on top of the magazine follower (with the bullet end toward the front of the magazine) and apply pressure with the thumb to fully seat the round in the magazine (fig. 25).

b. To load the magazine with a 5-round cartridge clip, the magazine filler is used (fig. 26). Slide the filler over the top rear portion of the magazine and insert a 5-round cartridge clip into the filler. Place either the thumb or the open end of the combination tool on the top round and push the 5 rounds into the magazine. Remove the clip and repeat the process until 20 rounds have been loaded into the magazine, then remove the magazine filler.

18. Loading the Magazine (in the Rifle)

a. To load a single round into an empty magazine in the weapon, lock the bolt to the rear and engage the safety. Place a round on top of the magazine follower and press down on the round and fully seat it in the magazine (fig. 27).

b. A magazine in the weapon can be loaded through the top of the receiver with a 5-round cartridge clip. To do this, place either end of the clip in the cartridge guide, then exert pressure with the thumb or the open end of the combination tool on the top round, forcing 5 rounds into the magazine (fig. 28). Remove and discard the cartridge clip. Repeat the process until the magazine is loaded.

19. Loading and Unloading the Rifle

a. Place the safety in the safe position.

b. Insert a loaded magazine into the magazine well, top front first, until the operating rod spring guide engages the magazine (1, fig. 29), then pull backward and upward until the magazine snaps into position (2, fig. 29). A click will be heard which indicates that the magazine is fully seated. Pull back and release the operating rod handle, allowing the bolt to strip the top round from the magazine and load it into the chamber.

c. Remove the magazine as described in paragraph 6.

Section II. FUNCTIONING

20. Semiautomatic

a. Each time a round is fired, the parts inside the rifle work together in a given order. *This is the cycle of operation.* This cycle is similar in all small arms. A knowledge of what happens inside the rifle during the cycle of operation will help you to understand the causes of, and remedies for, various stoppages.

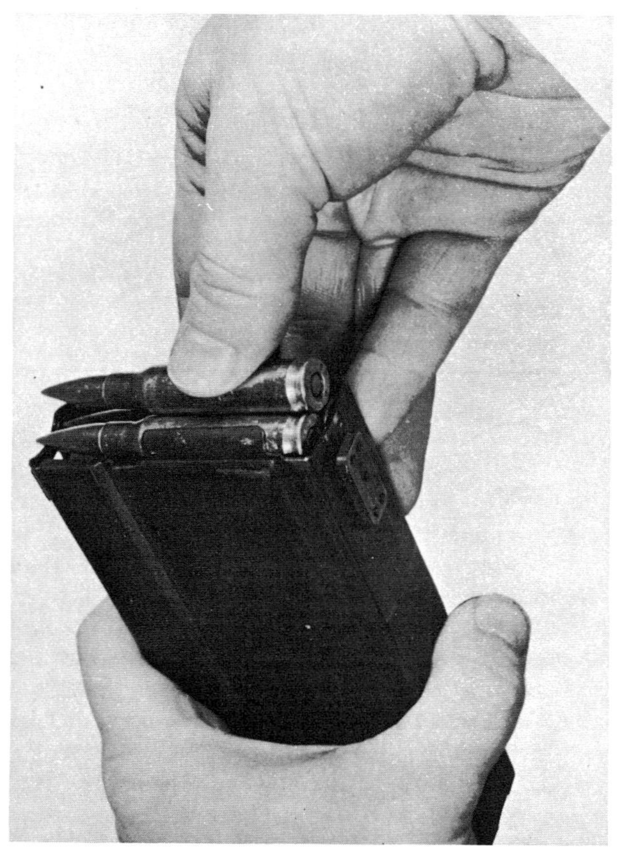

Figure 25. Loading the magazine single round (out of rifle).

Figure 26. Loading the magazine using the magazine filler (magazine out of rifle).

Figure 27. Loading the magazine with a single round (magazine in rifle).

Figure 28. Loading the magazine with a 5-round cartridge clip (magazine in rifle).

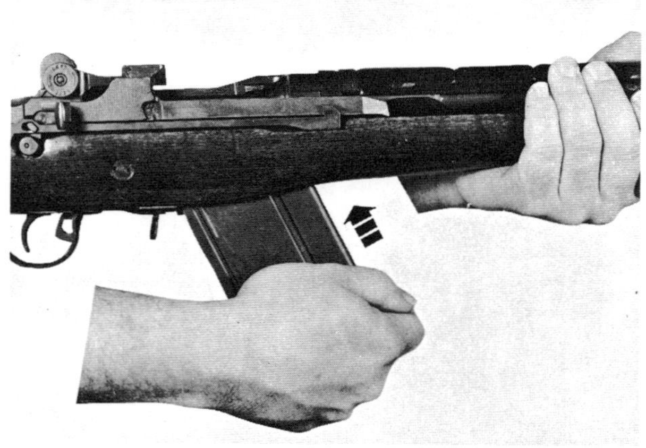

Figure 29. Loading the magazine into the rifle.

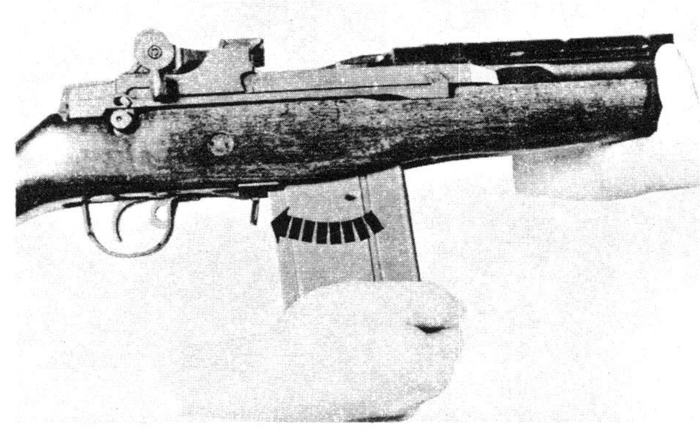

2

Figure 29—Continued.

b. The cycle of operation is broken down into eight steps. These steps are listed below, together with a brief description of what occurs inside the rifle during each step.

(1) *Feeding.* Feeding takes place when a round is forced into the path of the bolt. The top round is forced into the path of the bolt by the magazine follower which is under pressure of the magazine spring (fig. 30).

(2) *Chambering.* Chambering occurs when a round is moved into the chamber. This takes place as the bolt goes forward under pressure of the expanding operating rod spring, stripping the top round from the magazine and driving it forward into the chamber (fig. 31). Chambering is complete when the extractor snaps into the extracting groove on the cartridge and the ejector is forced into the face of the bolt.

(3) *Locking.* Locking begins as the bolt roller engages the camming surface in the hump of the operating rod. It is completed when the locking lugs of the bolt are fully seated in the locking recesses of the receiver (fig. 32).

(4) *Firing.* Firing occurs when the firing pin strikes the primer. As the trigger is pulled, the trigger lugs are disengaged from the hammer hooks and the hammer is released. The hammer moves forward under pressure of the hammer spring and strikes the tang of the firing pin, driving the firing pin against the primer, and firing the round (fig. 33).

(5) *Unlocking.* Unlocking (fig. 34) occurs after the firing of the round. As the bullet is forced through the barrel by the expanding gases, a small amount of gas enters the hollow gas piston, the gas cylinder, and the gas cylinder plug through the gas port. The expanding gases force the gas cylinder piston to the rear. It in turn drives the operating rod and bolt rearward. The operating rod cams the bolt roller upward, disengaging the locking lugs on the bolt from the locking recesses in the receiver. At this time the bolt is unlocked.

Note. The spindle valve must remain in the open position (the slot in the spindle head perpendicular to the barrel) during all firing, except when launching a grenade (fig. 35).

(6) *Extracting.* Extracting is pulling the empty cartridge from the chamber. Slow initial extraction takes place as the bolt unlocks. The bolt in its rearward motion pulls the empty cartridge with it (fig. 36).

(7) *Ejecting.* Ejecting is removing the empty cartridge from the receiver. As soon as the bolt has withdrawn the empty cartridge case clear of the chamber, the force of the ejector spring and plunger pushes the bottom edge of the cartridge base away from the bolt face, throwing it out and away from the receiver. When the last round has been fired, the bolt is held in the rearward position by the bolt lock.

(8) *Cocking.* Cocking is positioning the hammer so that it is ready to fire the next round. The bolt, as it moves to the rear, forces the hammer down and rides over it. The hammer is caught by the sear if the trigger is held to the rear and by the trigger lugs if the trigger has been released (fig. 37). In either case, the hammer is held in the cocked position.

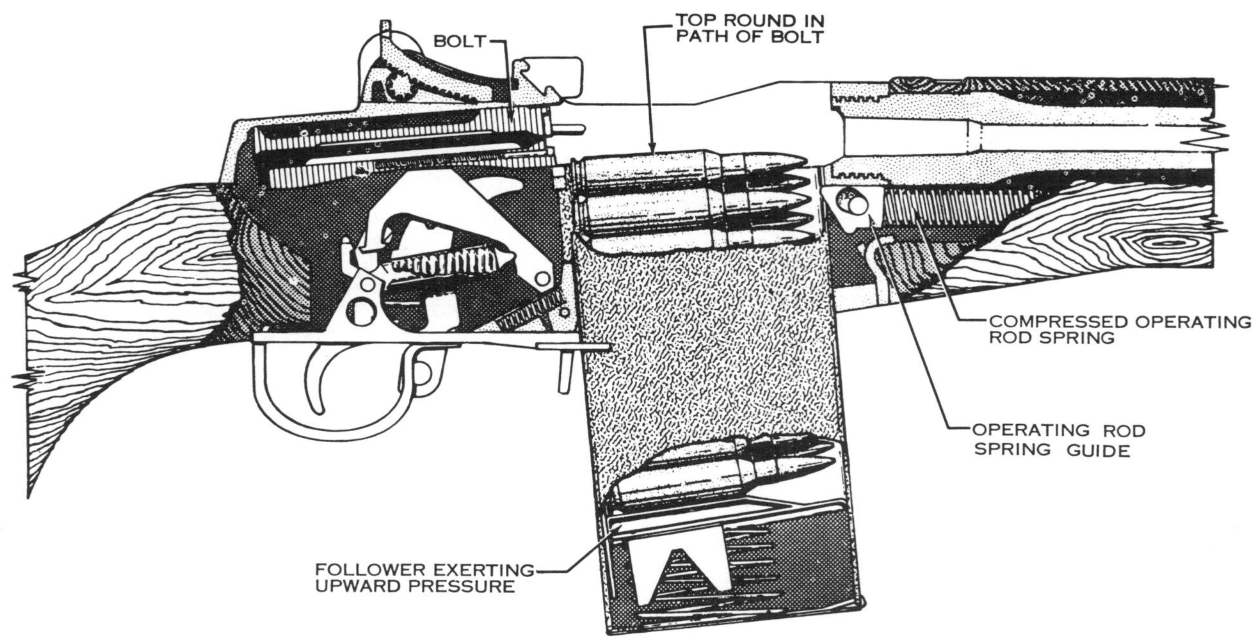

Figure 30. Feeding.

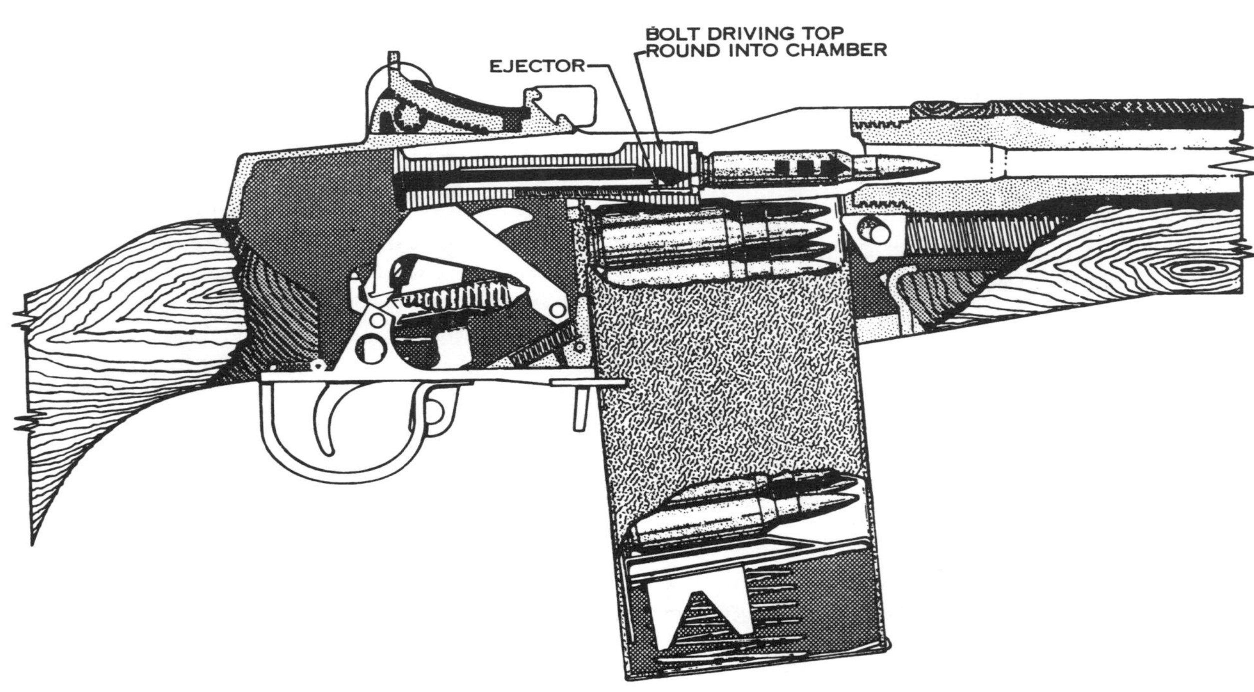

Figure 31. Chambering.

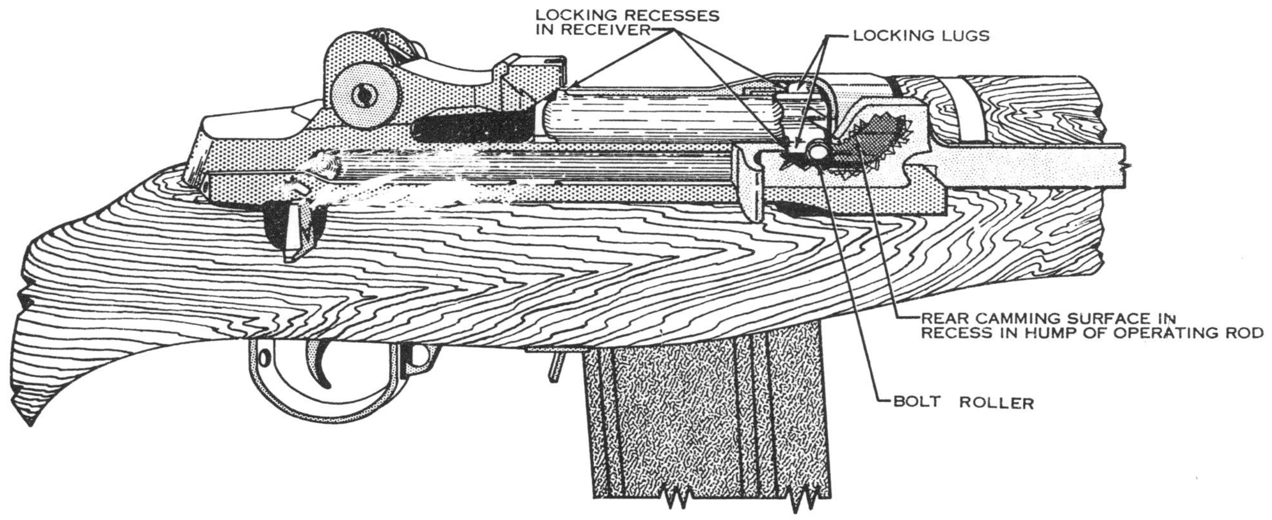

Figure 32. Locking.

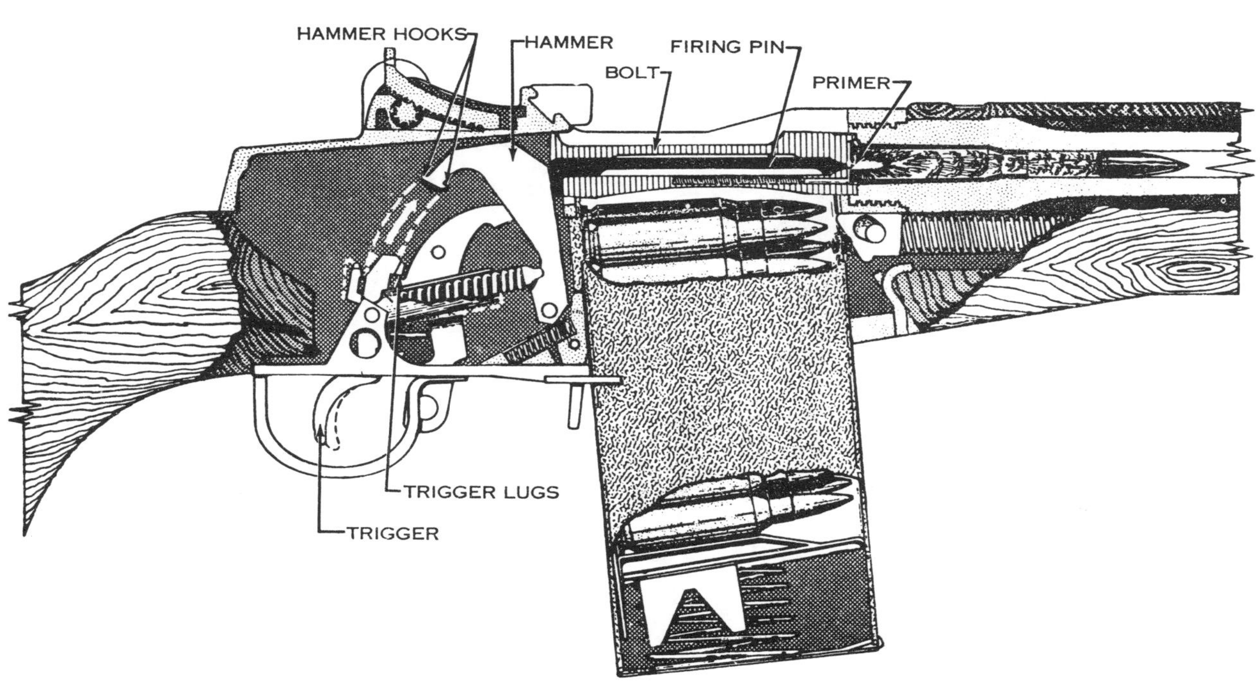

Figure 33. Firing.

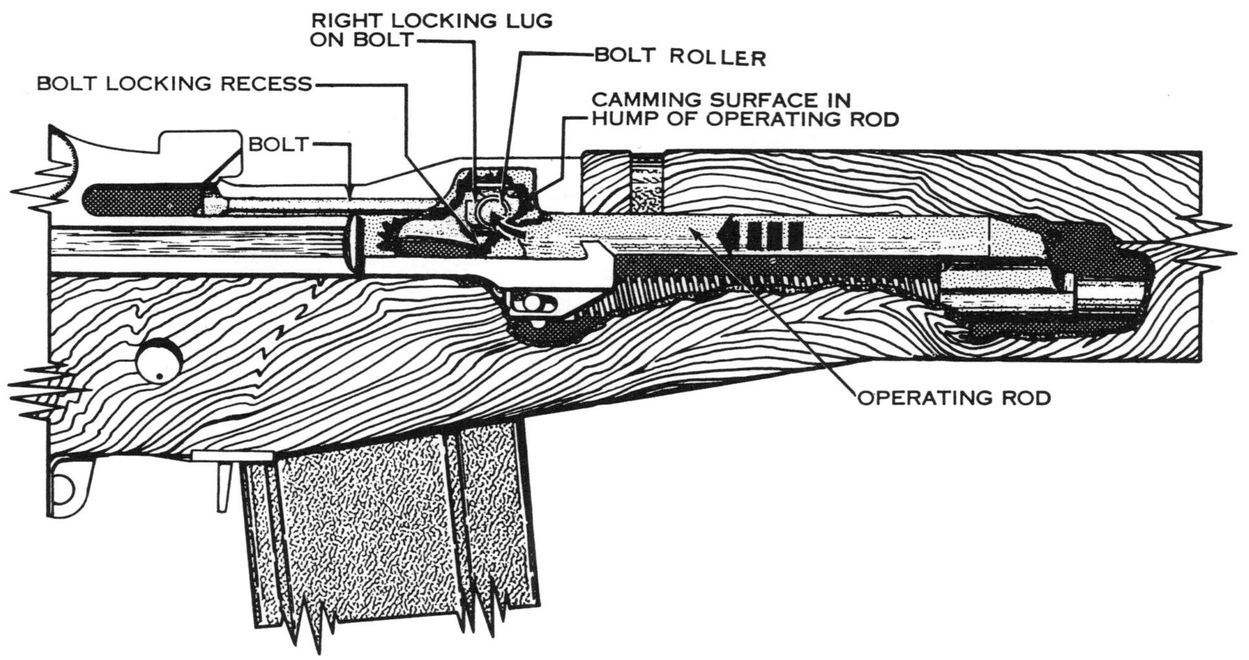

Figure 34. Unlocking.

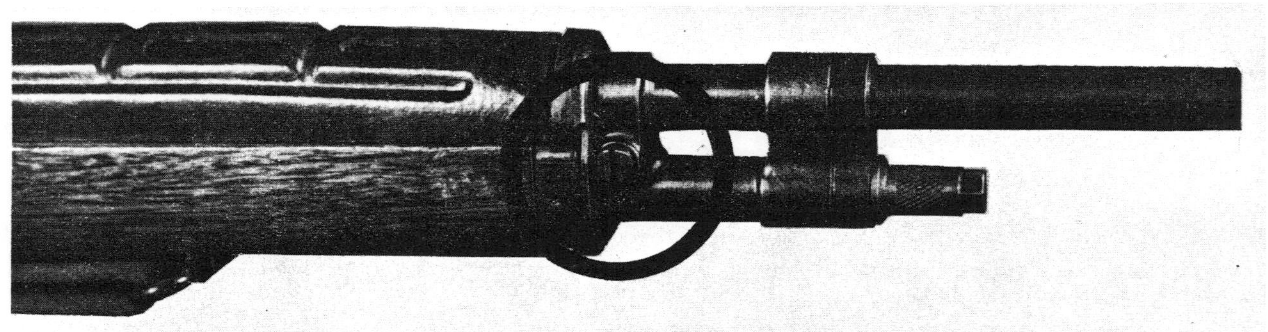

Figure 35. Positions of the spindle valve.

24

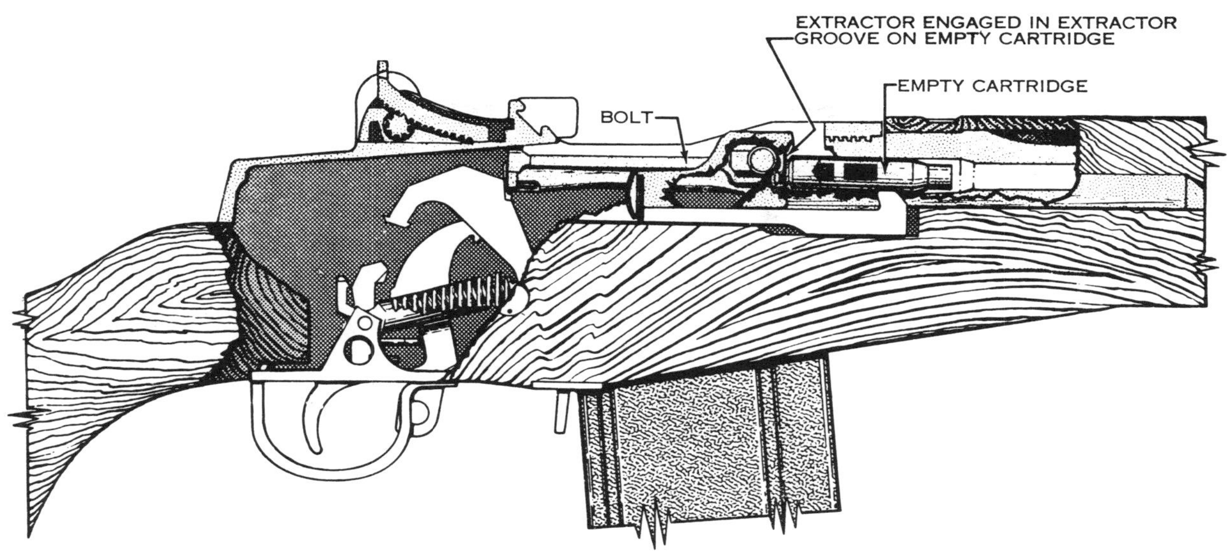

Figure 36. Extracting.

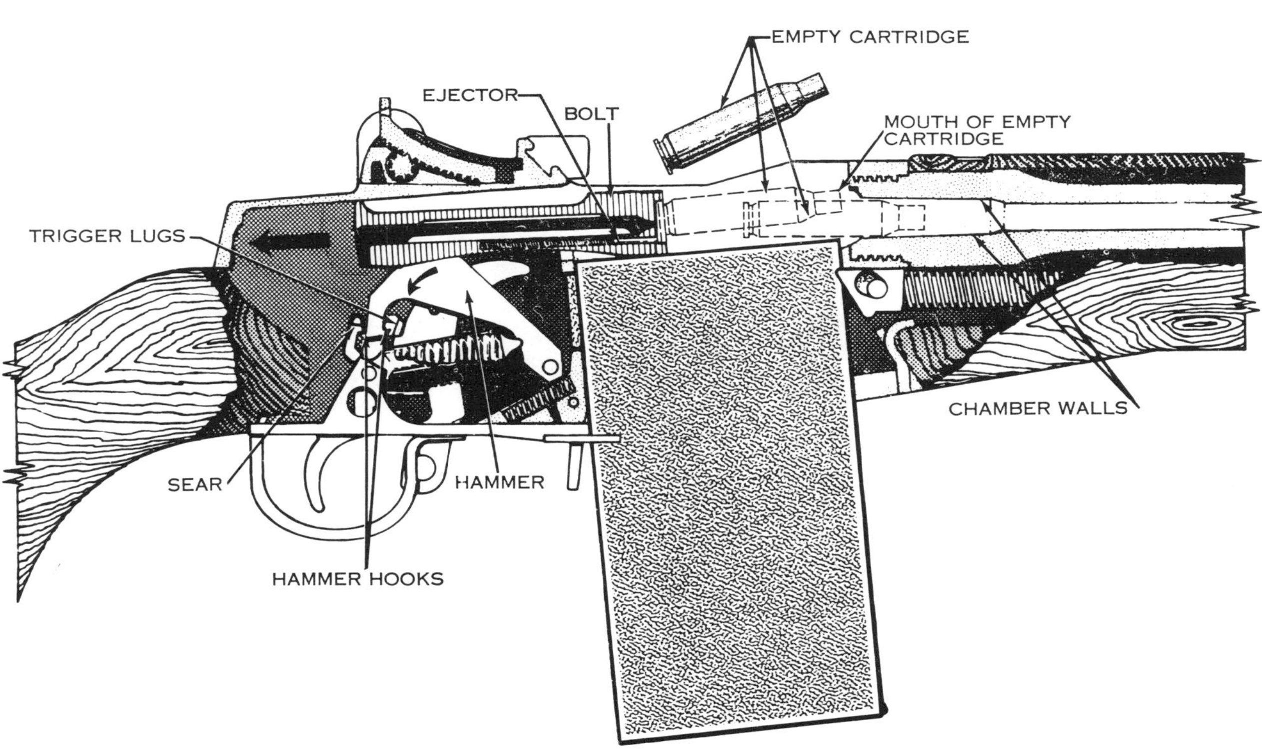

Figure 37. Ejecting the last round and cocking.

21. Automatic (Rifles Equipped With Selector)

a. When the selector is positioned with the face marked "A" to the rear (ear type projection up), the rifle is set for automatic fire. Turning the selector to automatic rotates the sear release in position to make contact with the sear.

b. After the first round has been fired (and with the trigger held to the rear), the operating rod starts its rearward movement under pressure of the expanding gases. As it moves to the rear, the connector assembly moves rearward under pressure of the connector assembly spring. The movement of the connector assembly rotates the sear release on the selector shaft so that the flange on the sear release allows the sear to move forward into a position where it can engage the rear hammer hooks (1, fig. 38). Then, when the bolt drives the hammer to the rear, the sear engages the rear hammer hooks and holds the hammer in the cocked position.

c. After the bolt moves forward and locks, the shoulder on the operating rod engages the hook of the connector assembly and forces it forward. This rotates the sear release on the selector shaft, causing the flange on the sear release to push the sear to the rear, disengaging it from the rear hammer hooks (2, fig. 38). The hammer will then go forward if the trigger is held to the rear. If the trigger is released at any time prior to the firing of the last round, the hammer will be held in the cocked position by the trigger lugs.

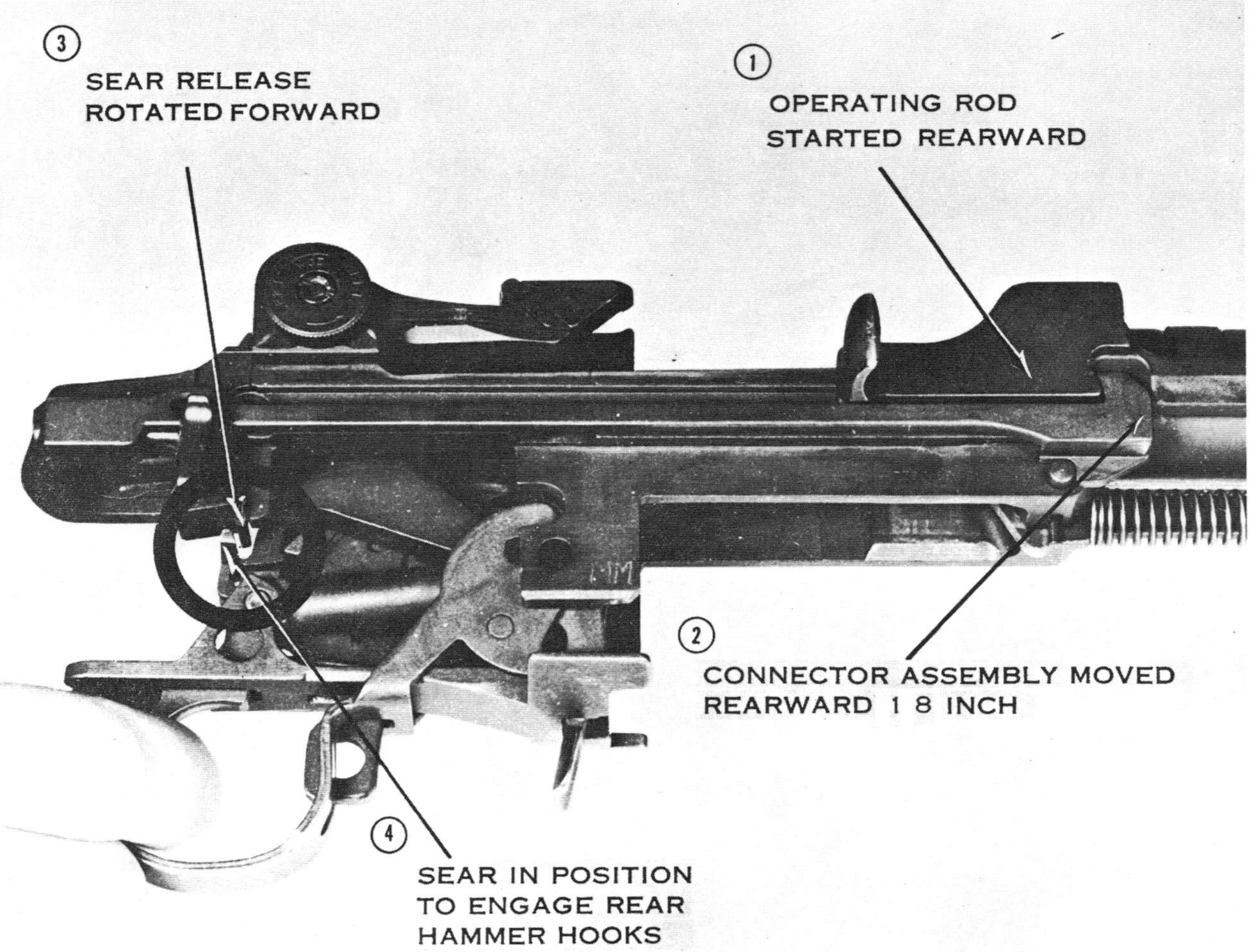

Figure 38. Actions of the connector assembly and its effects on the firing mechanism during automatic firing.

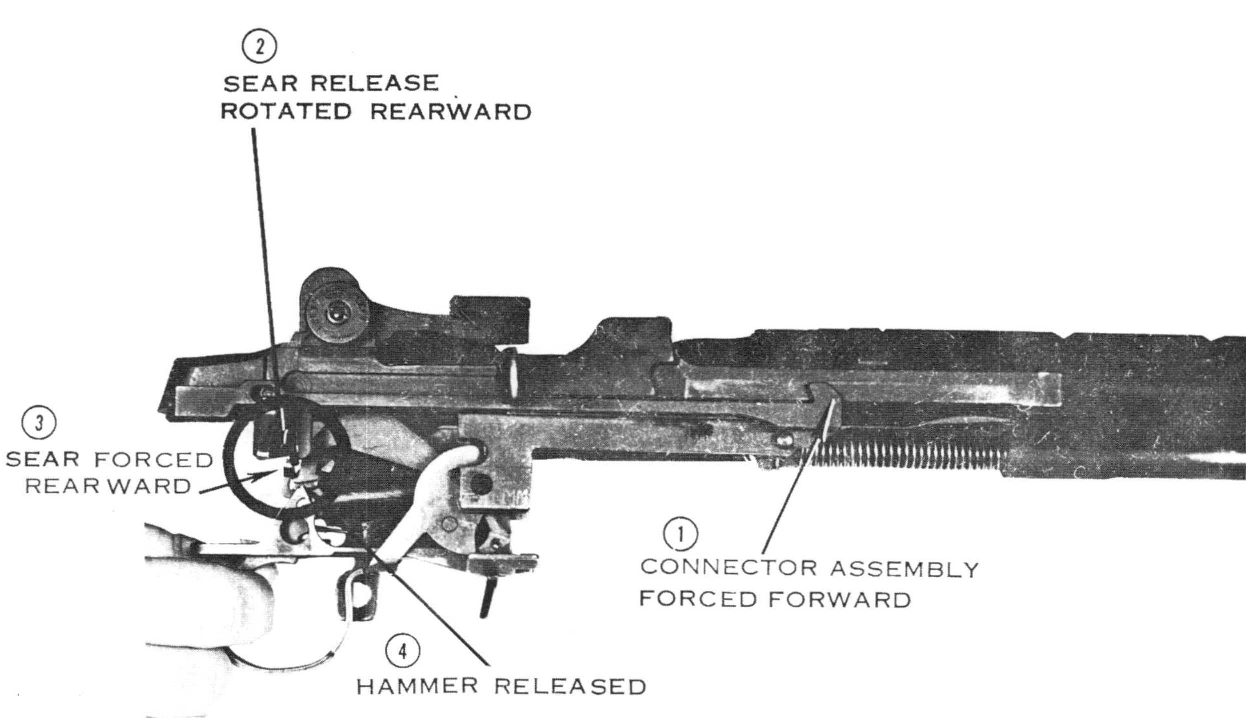

2

Figure 38—Continued.

*** * The instructions in Chapter 4 are written for the battlefield and are not recommended for civilian use. See the Appendix on page 49 for stoppage and immediate action instructions pertaining to civilian use. * ***

CHAPTER 4

STOPPAGES AND IMMEDIATE ACTION

22. Stoppages

a. A stoppage is any unintentional interruption in the cycle of operation. The stoppage may be caused by improper functioning of the rifle or faulty ammunition.

  b. Types of Stoppages.

  (1) *Misfire.* A misfire is a failure to fire. A misfire itself is not dangerous, but since it cannot be immediately distinguished from a delay in the functioning of the firing mechanism, or from a hangfire, it should be considered as a possible delay in firing until this possibility has been eliminated. A delay in the functioning of the firing mechanism could result from the presence of foreign matter such as sand, grit, oil and grease. These might create a partial mechanical restraint which, after some delay, is overcome by continued force applied by the spring, and the firing pin then striking the primer. No round should be left in a hot weapon any longer than necessary because of the possibility of a cookoff.

  (2) *Hangfire.* A hangfire is a delay in the functioning of a propelling charge at the time of firing. The amount of delay is unpredictable. A hangfire cannot be distinguished immediately from a misfire.

  (3) *Cookoff.* A cookoff is the functioning of a chambered round due to the heat of the weapon. If the primer or propelling charge should cookoff, the projectile will be propelled from the weapon with normal velocity even though no attempt was made to fire the primer by actuating the firing mechanism.

  c. Common Stoppages. The rifle will function efficiently if it is properly cared for. The firer must watch for defects and correct them before they cause a stoppage. Some of the more common stoppages, their usual causes, and remedies are shown in chart II.

Chart II. Stoppages: Their Causes and Remedies

Stoppages	Cause	Remedy
Failure to feed	Defective or worn parts	Replace parts.
	Dirty or dented magazine	Clean or replace magazine.
	Loose gas cylinder plug	Tighten plug.
Failure to chamber	Lack of lubrication of operating parts	Clean and lubricate parts.
	Defective ammunition	Replace ammunition.
	Dirty chamber	Clean chamber.
Failure to lock	Lack of lubrication of operating parts	Clean and lubricate parts.
	Dirty locking recesses	Clean recesses.
	Weak operating rod spring	Replace spring.
	Spindle valve closed	Open valve.
Failure to fire	Defective ammunition	Replace ammunition.
	Broken firing pin	Replace firing pin.
	Defective or broken parts in firing mechanism.	Replace parts or entire firing mechanism.
Failure to unlock	Dirty chamber	Clean chamber.
	Lack of lubrication of operating parts	Clean and lubricate parts.
	Insufficient gas	Tighten gas cylinder plug and check spindle valve.
Failure to extract	Dirty chamber	Clean chamber.
	Dirty ammunition	Replace ammunition.
	Broken extractor	Replace extractor.
Failure to eject	Broken ejector or weak ejector spring	Replace faulty part.
Failure to cock	Defective or broken parts in firing mechanism.	Replace parts or entire firing mechanism.

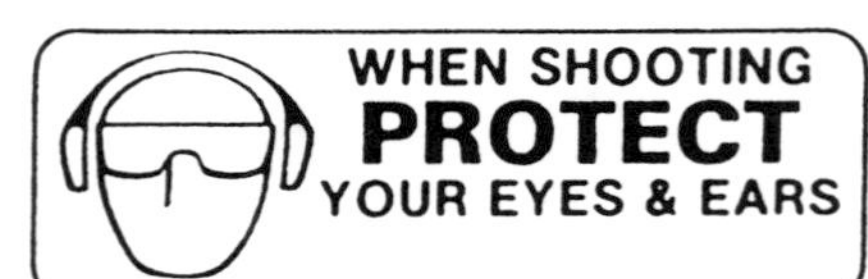

23. Immediate Action

Immediate action is the unhesitating application of a probable remedy to reduce a stoppage without investigating the cause. Immediate action is taught in two phases.

a. The first phase is taught as a drill so that the rifleman learns to perform it quickly and instinctively without thought as to the cause of the stoppage. To apply the first phase, with the right hand, palm up, pull the operating rod handle all the way to the rear. Release it, aim and attempt to fire. The palm is up to avoid injury to the hand in event of a cookoff or hangfire (fig. 39).

b. If the first phase of immediate action fails to reduce a stoppage, the second phase of immediate action is applied. The five key words used to help remember the steps in the second phase are: TAKE, PULL, LOOK, LOCATE, and REDUCE.

(1) TAKE the rifle from the shoulder.
(2) PULL the operating rod handle slowly to the rear.
(3) LOOK in the receiver.
(4) LOCATE the stoppage by observing, as the operating rod handle is pulled to the rear, what is in the chamber, and what has been ejected.
(5) REDUCE the stoppage and continue to fire.

c. Hangfires and misfires will occur rarely. Normally, the firer will instinctively apply immediate action which in most instances reduces the stoppage even when caused by a hangfire or misfire.

d. The normal cause of a misfire is faulty ammunition. Therefore, further use of ammunition from that lot should be suspended and reported to maintenance for disposition.

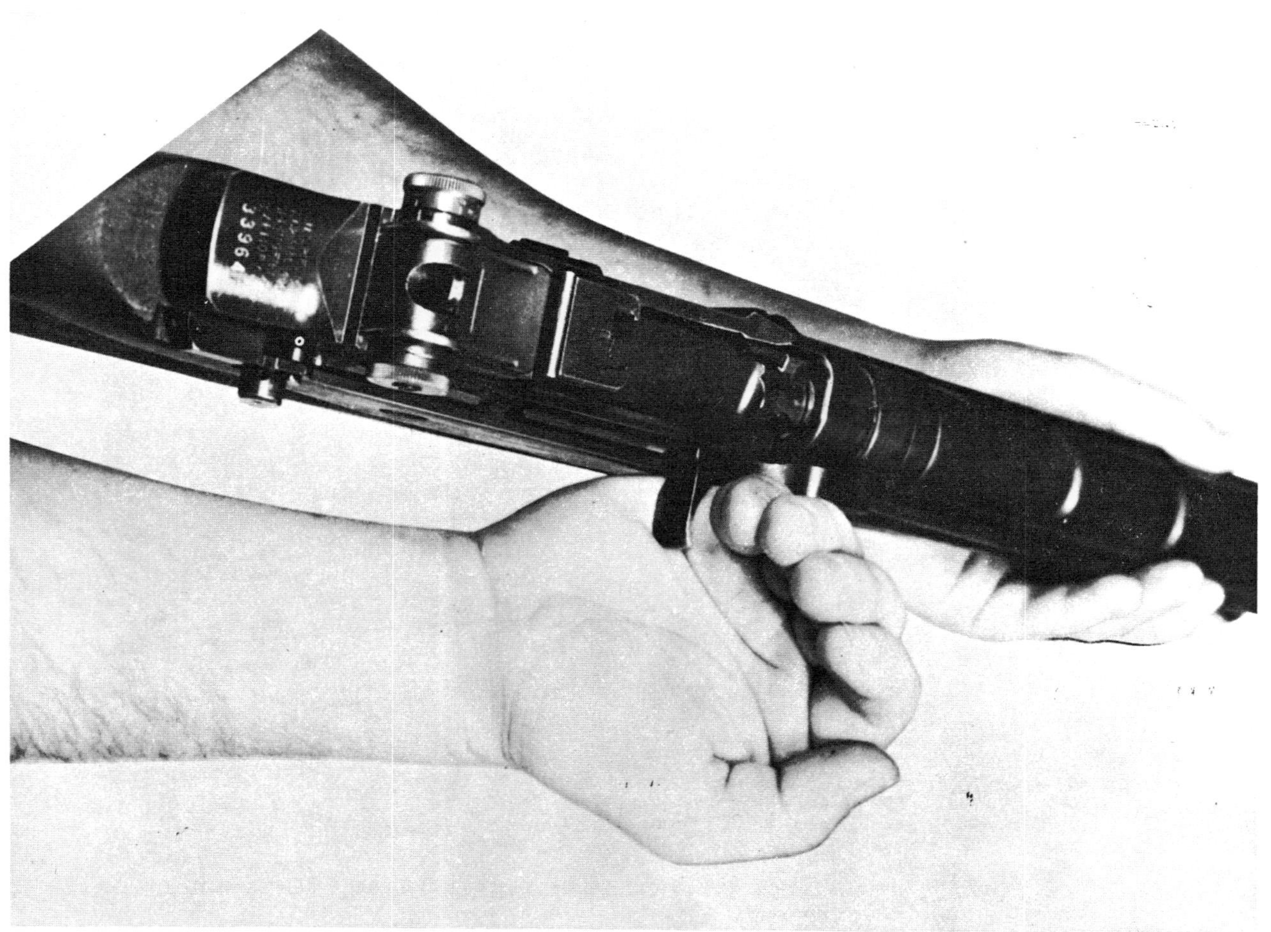

Figure 39. Applying immediate action.

24. General

Maintenance includes all measures taken to keep the rifle in operating condition. This includes normal cleaning, inspection for defective parts, repair, and lubrication.

25. Cleaning Materials, Lubricants, and Equipment

a. Cleaning Materials.

(1) Bore cleaner (cleaning compound solvent (CR)) is used primarily for cleaning the bore; however, it can be used on all metal parts for a temporary (1 day) protection from rust.

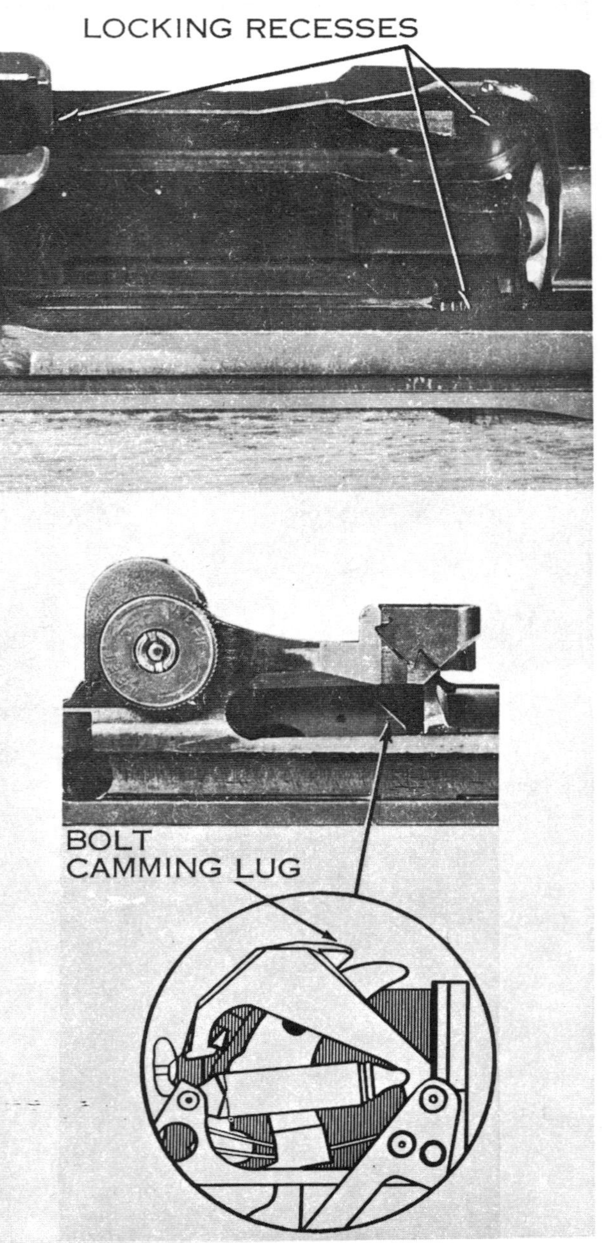

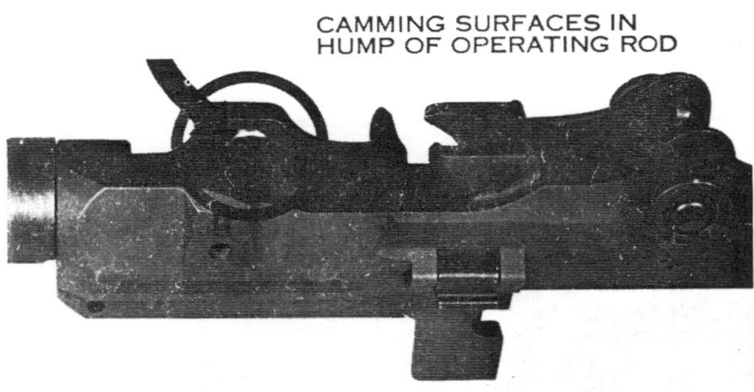

<table>
<tr><td>1</td><td>2</td></tr>
<tr><td>Figure 40. Points to apply rifle grease.</td><td>Figure 40—Continued.</td></tr>
</table>

(2) Hot, soapy water or plain hot water is no substitute for bore cleaner and *will only be used when bore cleaner is not available.*

(3) Drycleaning solvent is used for cleaning rifles which are coated with grease, oil, or corrosion-preventive compounds.

(4) Carbon-removing compound (PC111–A) is used on stubborn carbon deposits by soaking and brushing. This process must be followed by the use of drycleaning solvent.

b. Lubricants.

(1) Lubricating oil, general purpose, is used for lubricating the rifle at normal temperatures (PL special).

(2) Lubricating oil, weapons (LAW) is used for low temperatures (below 0°).

(3) OE 10 engine oil may be used as a field expedient under combat conditions when the oils prescribed in (1) and (2) above cannot be obtained. However, as soon as possible the weapon should be cleaned and lubricated with the proper, authorized lubricants.

(4) Rifle grease should be applied to those working surfaces shown in figure 40.

c. Equipment.

(1) A complete set of maintenance equipment (fig. 41) is stored in the stock of the M14 rifle and consists of a—

(*a*) Combination tool.

(*b*) Chamber cleaning brush.*

(*c*) Plastic case lubricant.

(*d*) Small arms cleaning rod case.

(*e*) Small arms cleaning rod section (4 each).

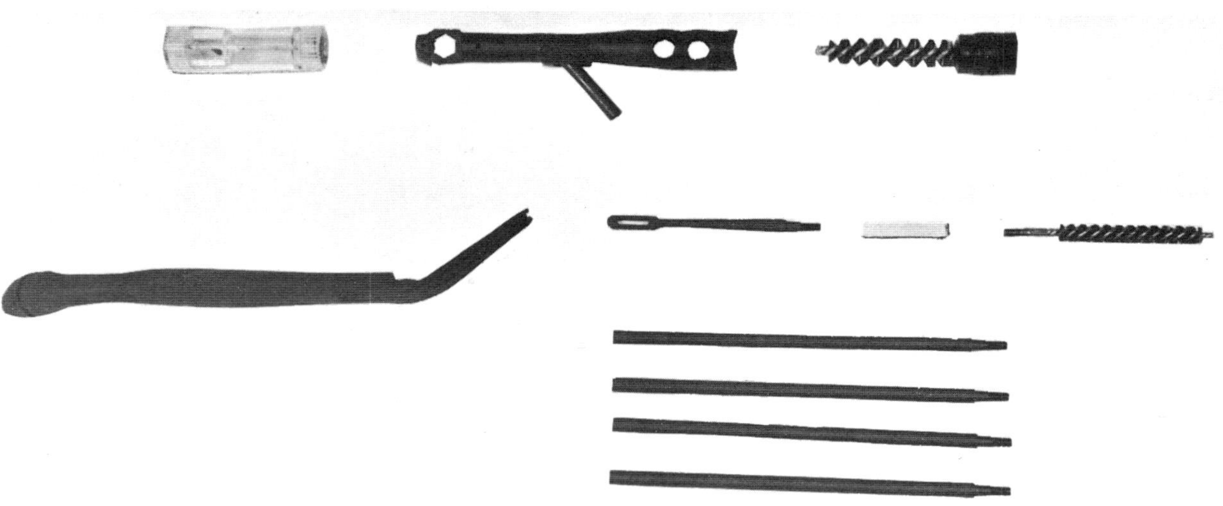

1

Figure 41. Maintenance equipment.

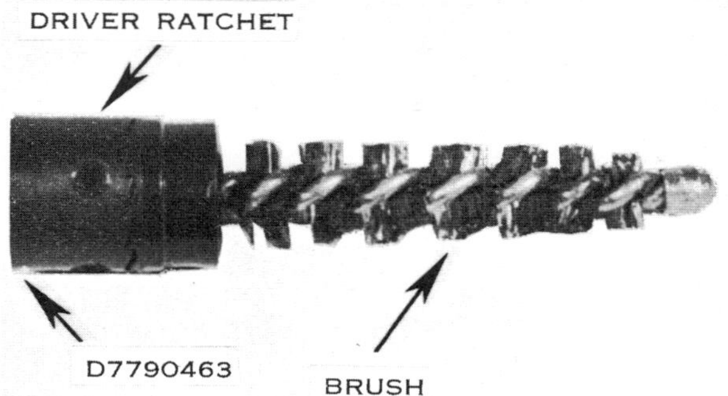

2

Figure 41—Continued.

(*f*) Cleaning patch holder.

(*g*) Small arms bore cleaning brush.

(2) The combination tool can be used as either a 20° offset screwdriver or as a gas plug wrench (figs. 42 and 43).

(*a*) The handle of the combination tool is also used as the cleaning rod handle. To do this, allow the cleaning rod extension of the tool to fall from the tool handle so that it hangs perpendicular.

---

*Insure the M14 chamber brush is used to prevent barrel damage. The M14 brush is one-half inch shorter than the M1 chamber brush.

Assemble the four sections of the cleaning rod and screw into the threaded hole in the cleaning rod extension. Either the bore brush or the cleaning patch holder may be attached to the end of the cleaning rod.

(b) The plastic lubricant case (fig. 44) is closed with a screw cap which has a stem (applicator) attached that is used to apply oil drop by drop on one end. The cap is fitted with a gasket to prevent oil leakage. The other end has another screw cap and contains rifle grease.

26. Cleaning the Rifle

a. Procedures for Cleaning Chamber and Bore. The rifle must be cleaned after it has been fired because firing deposits primer fouling, powder ashes, carbon and metal fouling. The ammunition has a noncorrosive primer which makes cleaning easier, but not less important. The primer still leaves a deposit that may collect moisture and promote rust if it is not removed. The procedures for cleaning the chamber and bore are described in figures 45 and 46. These procedures will insure that the bore is cleaned evenly and will prevent foreign matter from being pushed into the receiver from the bore. Upon completion of firing, bore cleaner should be applied for ease of further cleaning.

b. Gas Cylinder Plug. Pour a small quantity of bore cleaner in the plug, insert and rotate the bore cleaning brush. Remove the brush, clean and dry the plug with patches.

c. Gas Cylinder. Install the patch holder on a section of the cleaning rod. Put two patches in the holder, moisten them with bore cleaner and swab the cylinder bore. *Dry the cylinder bore with clean patches. Use no abrasives in cleaning the cylinder and do not oil the interior surfaces.*

d. Gas Piston. Saturate patches with bore cleaner and wipe the exterior surface of the piston

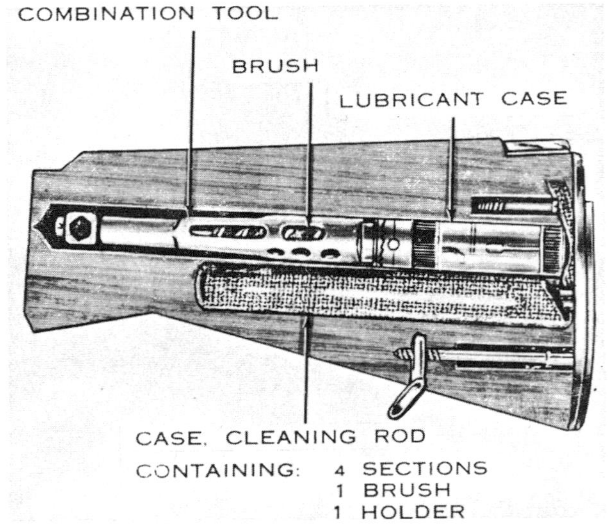

Figure 41—Continued.

Figure 42. Combination tool used as a screwdriver.

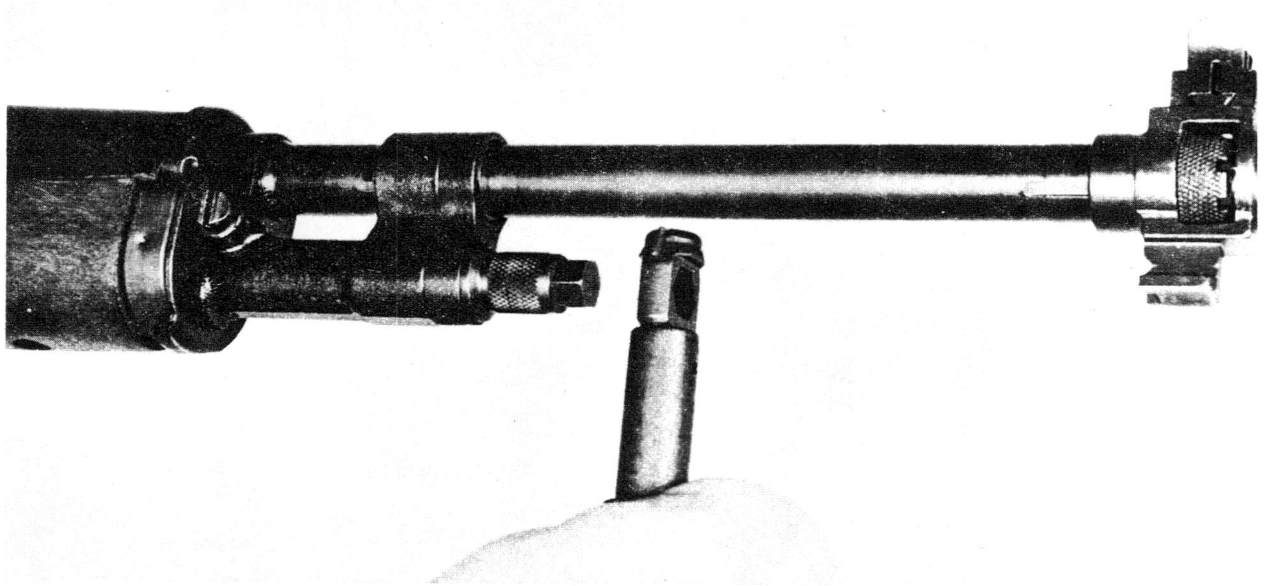

Figure 43. Combination tool used as a wrench.

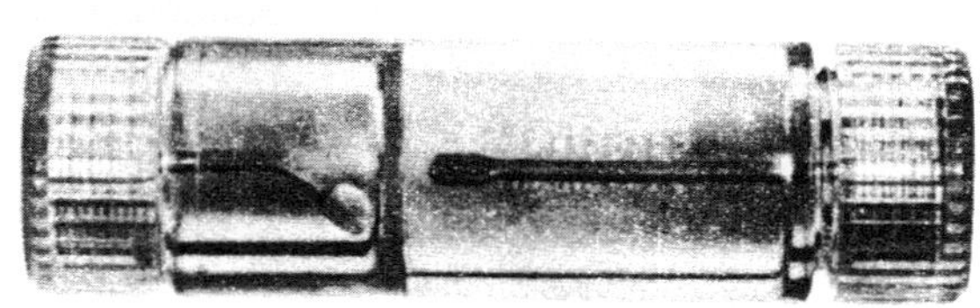

Figure 44. Plastic lubricant case.

as clean as possible. Install the bore cleaning brush on a section of the cleaning rod. Moisten the brush with bore cleaner and clean the interior of the piston. Wipe the piston dry, *but do not oil.* The gas system incorporates a self-cleaning section and functions within very close tolerances. A piston does not have to be shiny to function properly. *Do not use abrasives to clean the piston.*

e. Face of the Bolt. Clean the face of the bolt with a patch and bore cleaner, paying particular attention to its inside edges. Remove the bore cleaner with dry patches and oil the part lightly.

f. Spindle Valve. Depress the valve and rotate it several times after each day's firing. Do not disassemble it.

g. Magazine. Inspect the interior of the magazine by depressing the follower with the thumb. If the interior is dirty, disassemble the magazine and clean it, then lightly oil the component parts.

Otherwise, merely wipe the magazine assembly clean and dry, then oil it.

h. Stabilizer Assembly. The stabilizer assembly should be removed and cleaned with a stiff brush to remove all carbon or other particles which may block the gas ports.

i. All Other Parts. Use a dry cloth to remove all dirt or sand from other parts and exterior surfaces. Apply a light coat of oil to the metal parts and rub *raw linseed oil* into the wooden parts. Care must be taken to prevent linseed oil from getting on metal parts.

j. After Firing. The rifle must be thoroughly cleaned no later than the evening of the day it is fired. For three consecutive days thereafter, check for evidence of fouling by running a clean patch through the bore and inspecting it. The bore should be lightly oiled after each inspection.

27. Normal Maintenance

a. The rifle should be inspected daily, when in use, for evidence of rust and general appearance. A light coat of oil should be maintained on all metal parts, except the gas piston, interior of the gas cylinder, and the gas plug.

b. The daily inspection should also reveal any defects such as burred, worn or cracked parts. Defects should be reported to the armorer for correction.

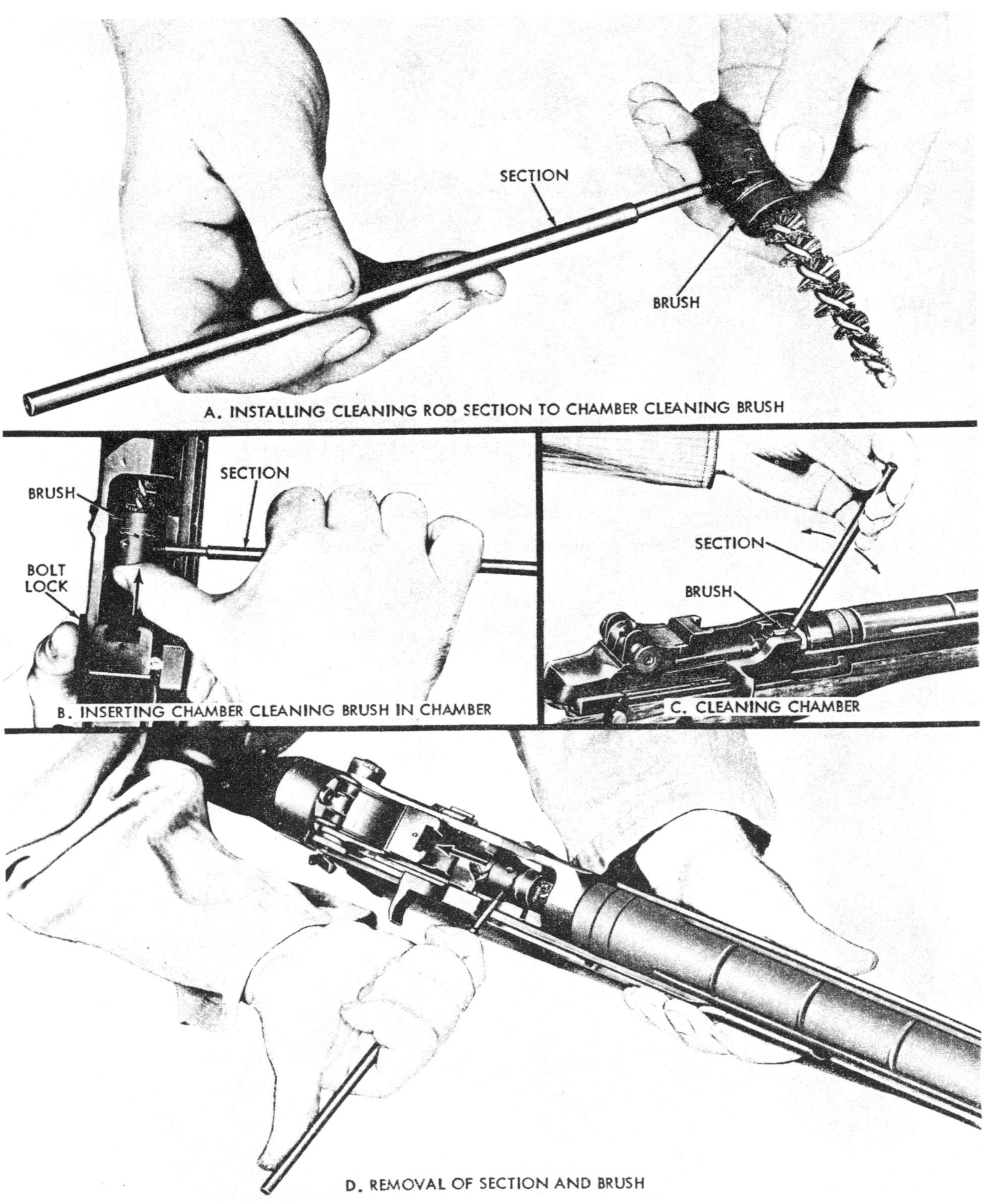

Figure 45. Cleaning the chamber.

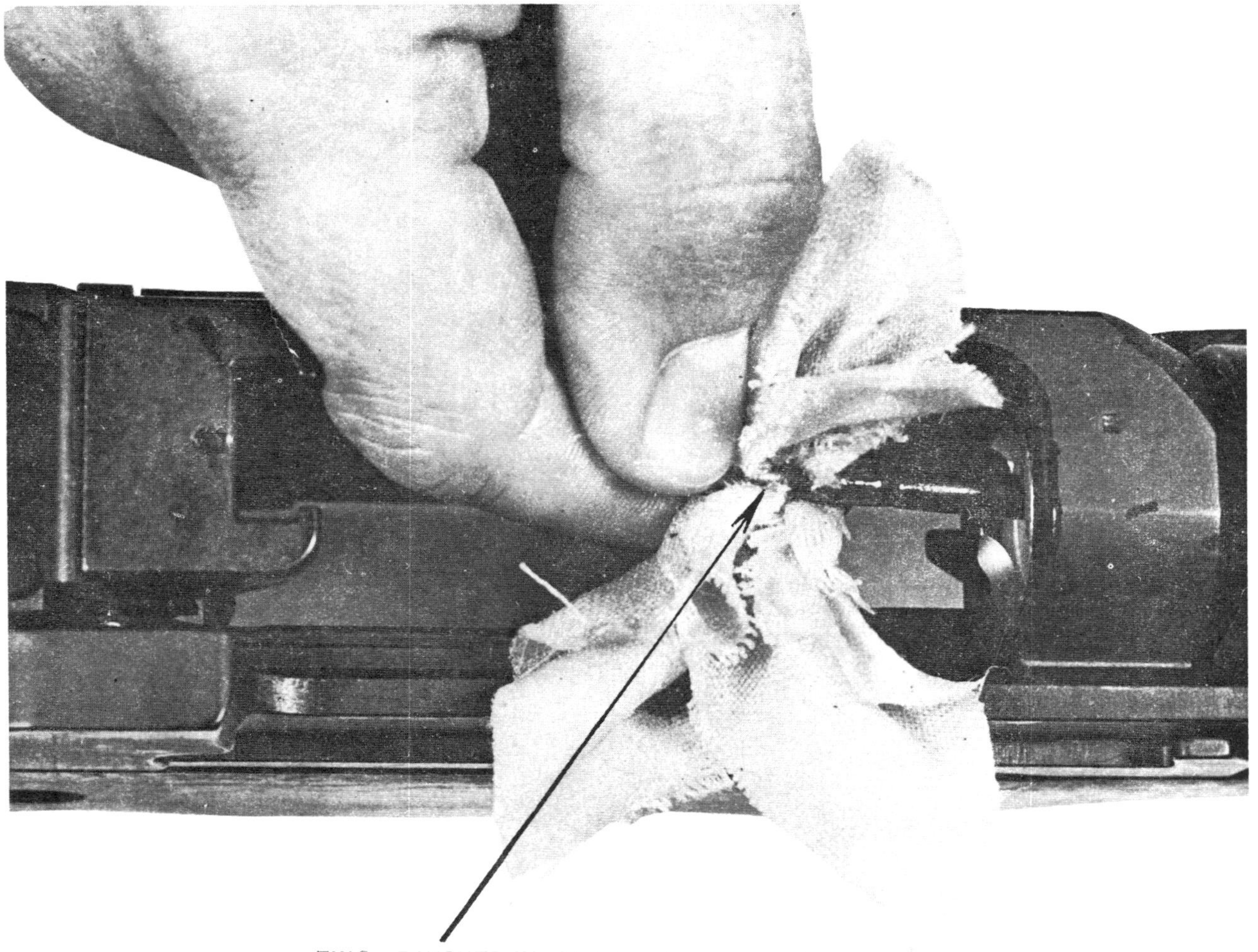

SCREW THE JOINTED CLEANING ROD TOGETHER
FIRMLY (LESS THE PATCH HOLDER) GENTLY INSERT
IT INTO THE BORE ALL THE WAY. (AN AUTHORIZED
SOLID NONJOINTED ROD MAY BE USED). FLARE THE
PATCHES OUT THEN INSERT THE PATCH HOLDER WITH
WET PATCHES INTO THE CHAMBER. PUSH THE
THREADED END INTO THE CHAMBER UNTIL IT
TOUCHES THE CLEANING ROD. HOLD IT THERE
WITH ONE HAND.

2

Figure 45—Continued.

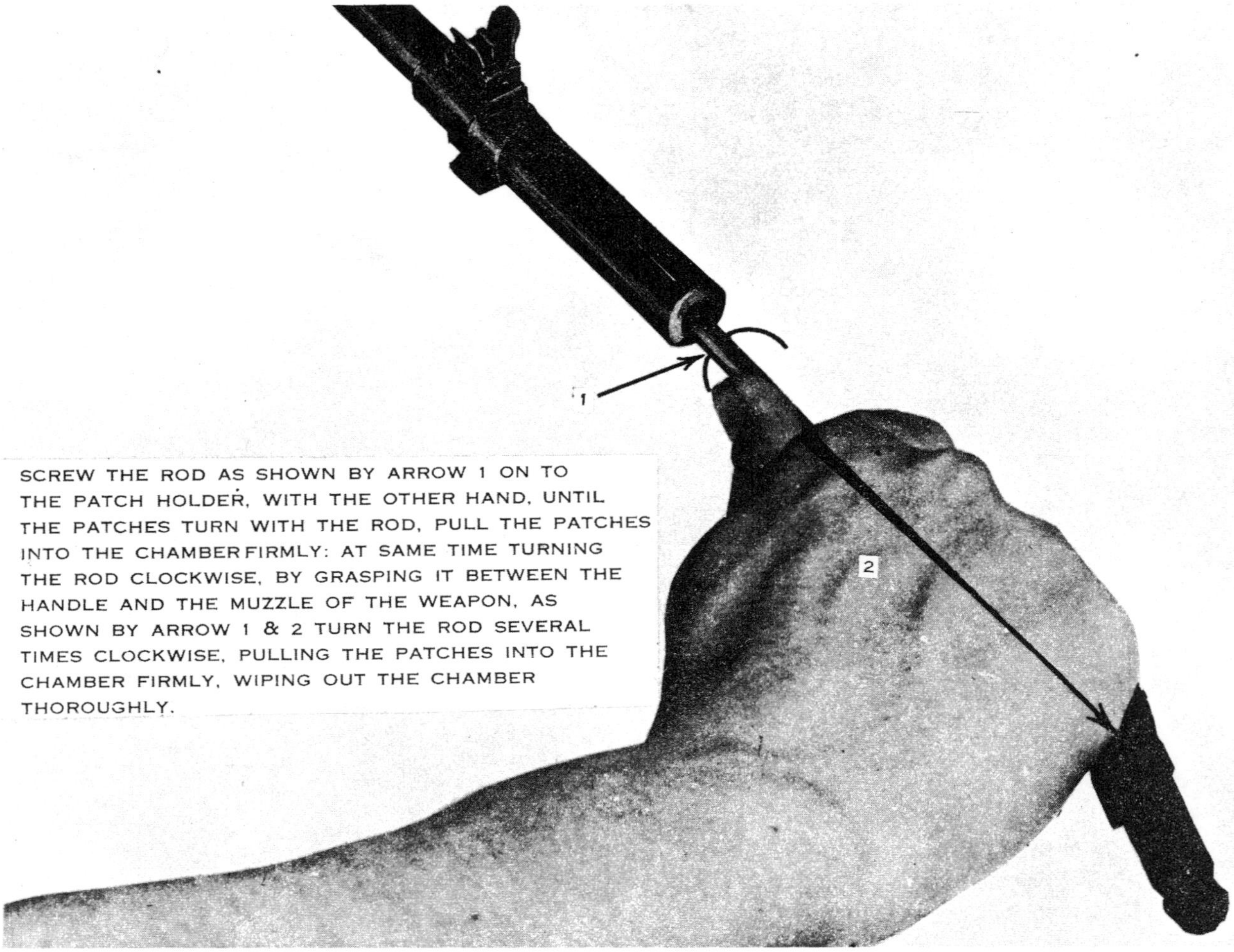

3

Figure 45—Continued.

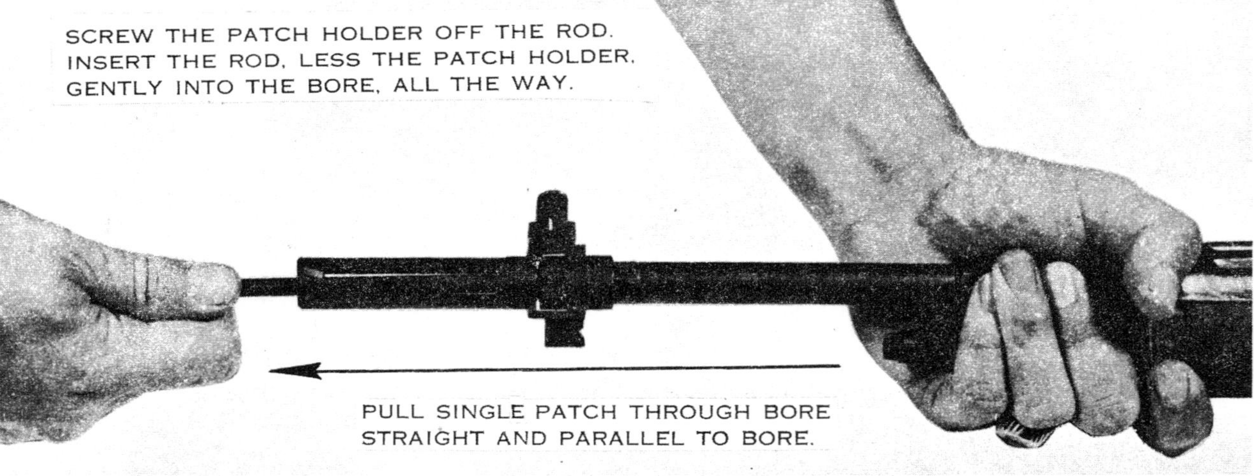

1

Figure 46. Cleaning the bore.

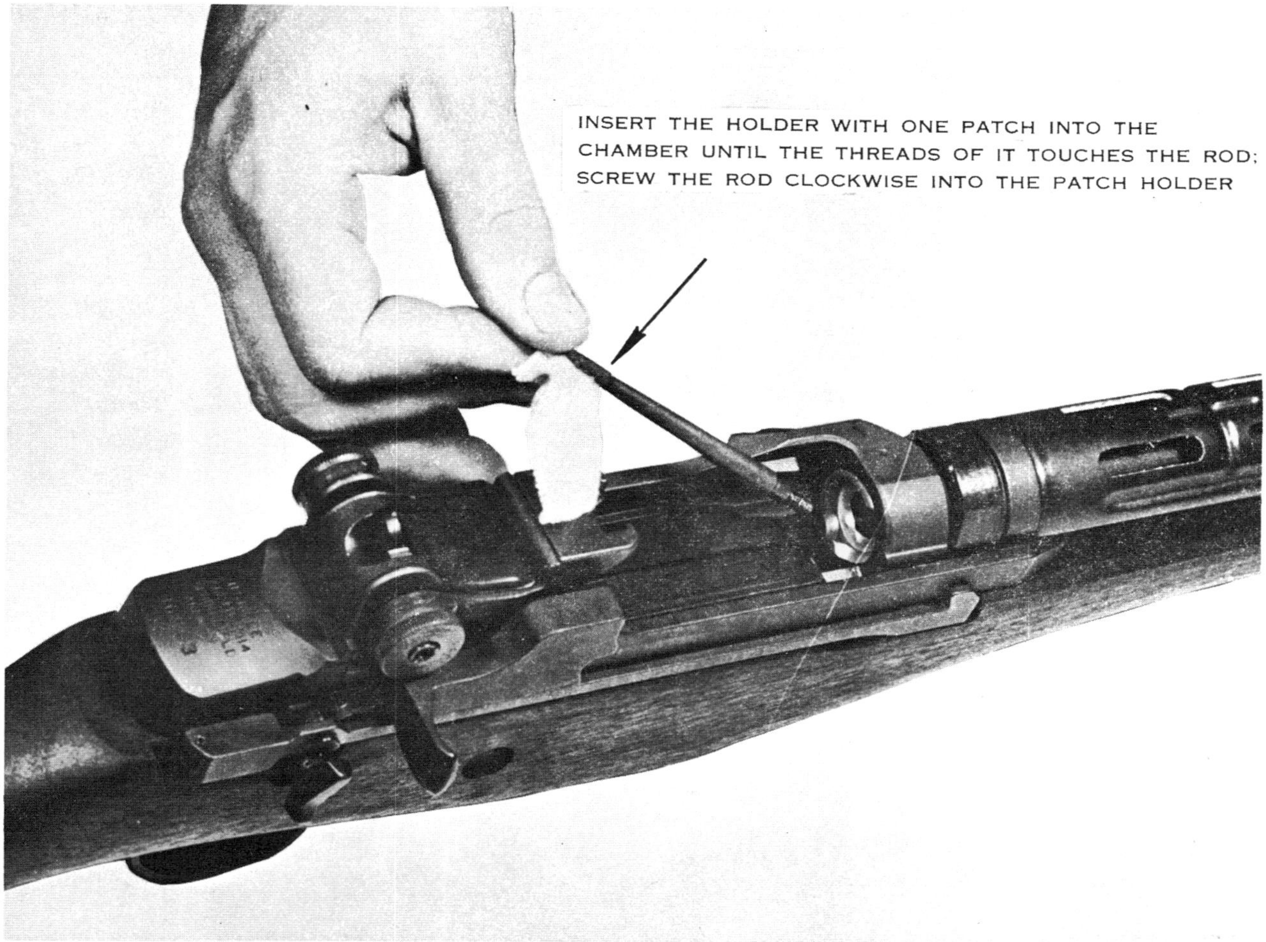

2

Figure 46—Continued.

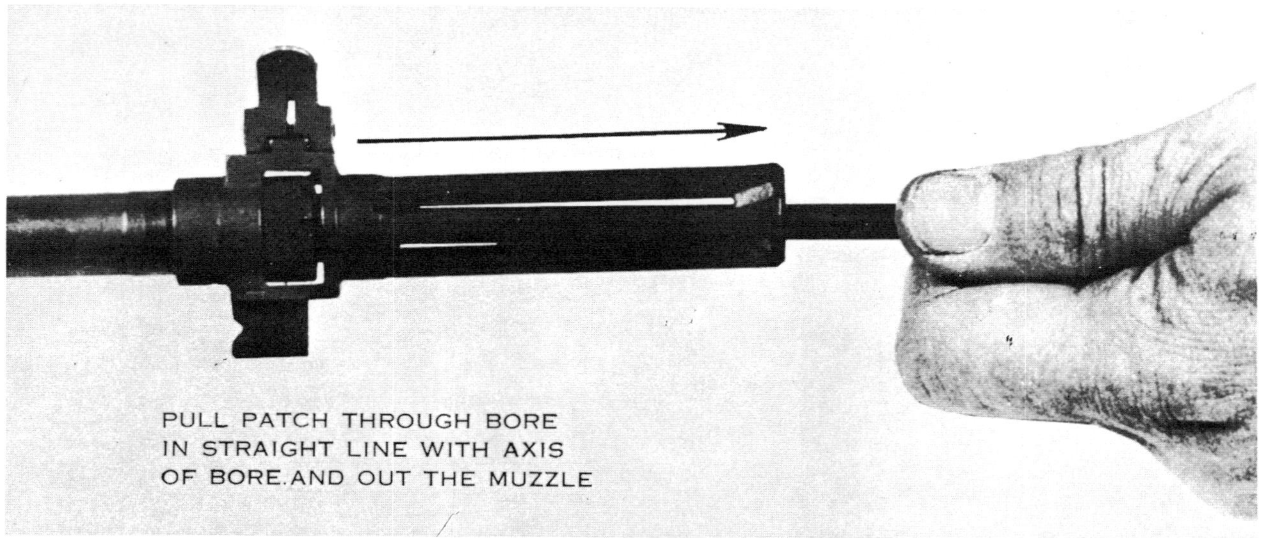

3

Figure 46—Continued.

c. A muzzle plug should never be used on the rifle. It causes moisture to collect in the bore forming rust and creating a safety hazard.

d. Obtaining the proper rear sight tension is extremely important; without it, the sight will not hold its adjustment in elevation. During normal maintenance, and prior to firing, the rear sight must be checked for correct sight tension. The indications of improper sight tension are:

(1) Elevation knob extremely difficult to turn.

(2) Elevation knob turns freely without an audible click.

(*a*) If the elevation knob is extremely difficult to turn, rotate the *windage* knob nut counterclockwise *one click at a time* with the screwdriver portion of the combination tool. After each click attempt to turn the *elevation* knob. Repeat this process until the *elevation* knob can be turned without extreme difficulty (1, fig. 47).

(*b*) If the elevation knob is extremely loose and the rear sight aperture will not raise, the *windage* knob nut must be turned in a *clockwise* direction, one click at a time, until the aperture can be raised

(*c*) To check for proper tension, the procedures listed below should be followed:

1. Raise the aperture to its full height.

2. Lower the aperture two clicks.

3. Grasp the rifle with the fingers around the small of the stock and exert downward pressure on the aperture with the thumb of the same hand (2, fig. 47).

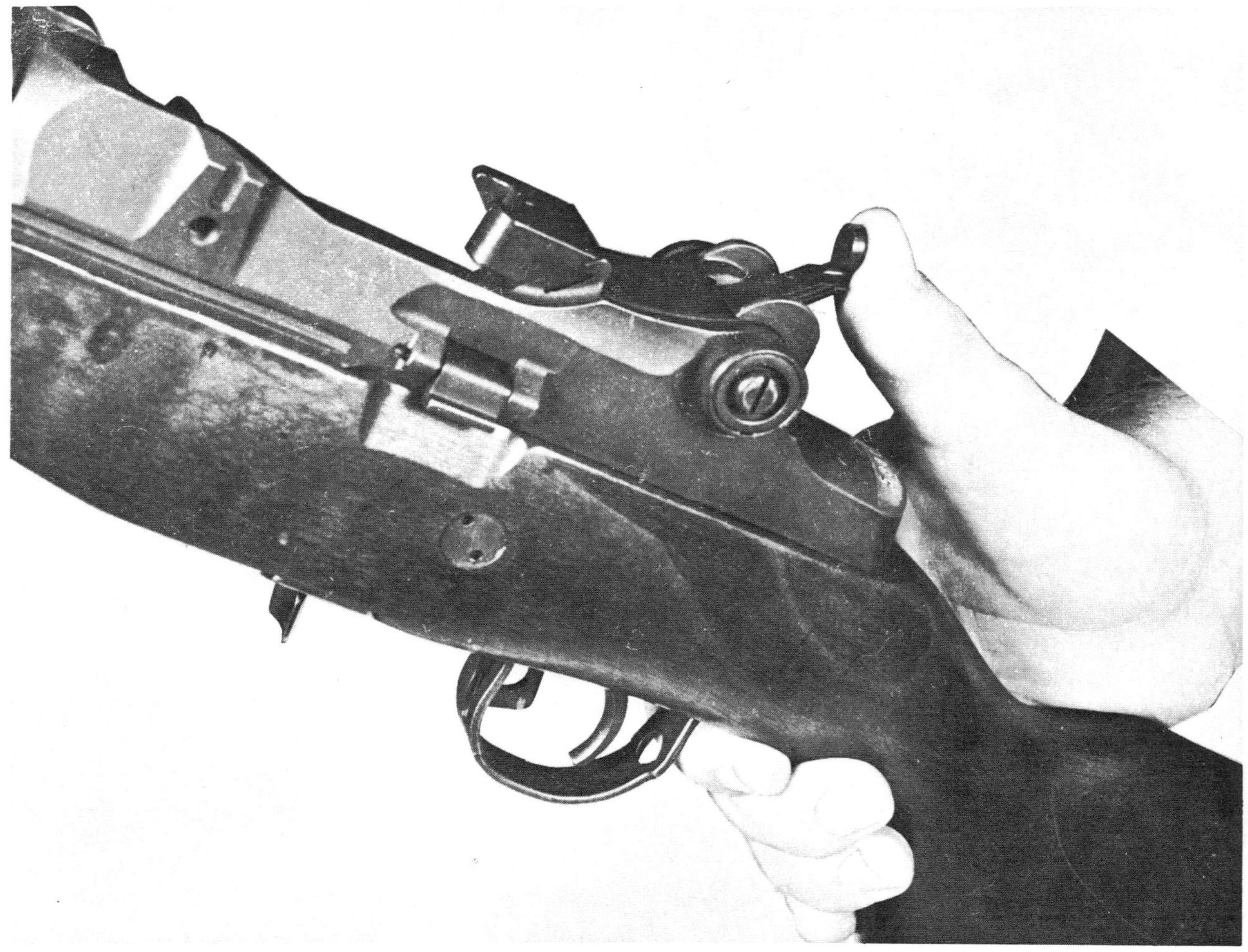

Figure 47. Adjusting sight tension.

(*d*) If the aperture drops, sight tension must be adjusted. To do this, the *windage knob nut* must be turned in a clockwise direction, one click at a time, until the aperture can no longer be pushed down as indicated in (*c*)*3* above. If the proper tension cannot be obtained, the rifle must be turned in to the unit armorer.

28. Special Maintenance

a. Before firing the rifle, the bore and the chamber should be cleaned and dried. A light coat of oil should be placed on all other metal parts, except those which come in contact with ammunition, the gas piston, interior of the gas cylinder, and the gas plug.

b. Before firing, rifle grease should be applied to the parts indicated in figure 40. A small amount of grease is taken up on the stem of the grease container cap and is applied at each place. Rifle grease is *not* used in extremely cold temperatures or when the rifle is exposed to extremes of sand and dust.

c. In cold climates (temperatures below freezing) the rifle must be kept free of moisture and excess oil. Moisture and excess oil on the working parts cause them to operate sluggishly or fail completely. The rifle must be disassembled and wiped with a clean dry cloth. Drycleaning solvent may be used if necessary to remove oil or grease. Parts that show signs of wear may be wiped with a patch lightly dampened with lubricating oil (LAW). It is best to keep the rifle as close as possible to outside temperatures at all times to prevent the collection of moisture which occurs when cold metal comes in contact with warm air. When the rifle is brought into a warm room, it should not be cleaned until it has reached room temperature.

d. In hot, humid climates or if exposed to salt water or salt water atmosphere, the rifle must be inspected thoroughly each day for moisture and rust. It should be kept lightly oiled with special preservative lubricating oil. *Raw linseed oil* should be frequently applied to the wooden parts to prevent swelling.

e. In hot, dry climates, the rifle must be cleaned daily or more often to remove sand and/or dust from the bore and working parts. In sandy areas, the rifle should be kept dry. The muzzle and receiver should be kept covered during sand and dust storms. Wooden parts must be kept oiled with *raw linseed oil* to prevent drying. The rifle should be lightly oiled when sand or dust conditions decrease.

f. Special instructions on caring for the rifle when it is subject to nuclear, biological or chemical contamination can be found in TM 3–220 and FM 21–40.

2

Figure 47—Continued.

29. General

The M14 rifle fires several types of ammunition. The rifleman should be able to recognize them and know which type is best for certain targets. He should also know how to care for the ammunition.

a. Figure 48 shows the parts of a typical cartridge.

b. The term "bullet" refers only to a small arms projectile; the term "ball" was originally used to describe the ball-shaped bullet of very early small arms ammunition. The term "ball ammunition" now refers to a cartridge with a general purpose solid core bullet intended for use against personnel and material targets.

30. Description

The types of ammunition can be identified by their individual markings (fig. 49).

a. Armor Piercing. The M61 armor piercing cartridge is used against lightly armored targets. The cartridge can be identified by its black tip.

b. Ball. The three types of ball ammunition (M59, M80 and M198 duplex) are used against personnel and unarmored targets. The M59 and M80 cartridges can be identified by their unpainted tips. The M198 duplex round can be identified by its green tip.

c. Tracer. The M62 tracer cartridge is used for indicating target areas and adjusting fire. The cartridge can be identified by its orange tip.

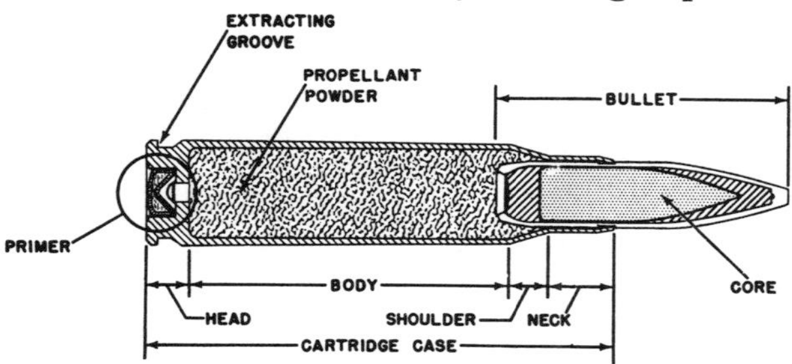

Figure 48. Parts of a cartridge.

d. Grenade Cartridge. The M64 rifle grenade cartridge is used for launching grenades and pyrotechnics. The cartridge can be identified by its five-pointed, star-crimped end.

e. Blank. The M82 blank cartridge is used to aid realism in training. It can be identified by its long narrow neck.

f. Dummy. The M63 dummy cartridge has six longitudinal corrugations approximately one-third the length of the case. There are no markings on the bullet and there is no primer in the base of the cartridge. It is used in training for dry firing exercises.

31. Packaging

a. 5-Round Cartridge Clip. Ammunition is prepacked in 5-round cartridge clips. Twelve clips are packed in a cloth bandoleer. Seven bandoleers are packed in a can and two cans are packed in a case.

b. 20-Round Carton. Ammunition is also packed in 20-round cartons. Twenty-three cartons are packed in a can and two cans are packed in a case.

c. Magazine Filler. The magazine filler is an adapter which fits over the top of an empty magazine (when the magazine is *not* in the weapon) and makes it easier to load. One magazine filler is packed in each case of ammunition.

32. Care, Handling, and Preservation

a. Care should be taken to prevent ammunition boxes from becoming broken or damaged.

b. Ammunition should not be exposed to the direct rays of the sun. If the powder is heated, excessive pressure may develop. This condition will affect ammunition performance and creates a safety hazard.

c. Ammunition should be kept clean and dry.

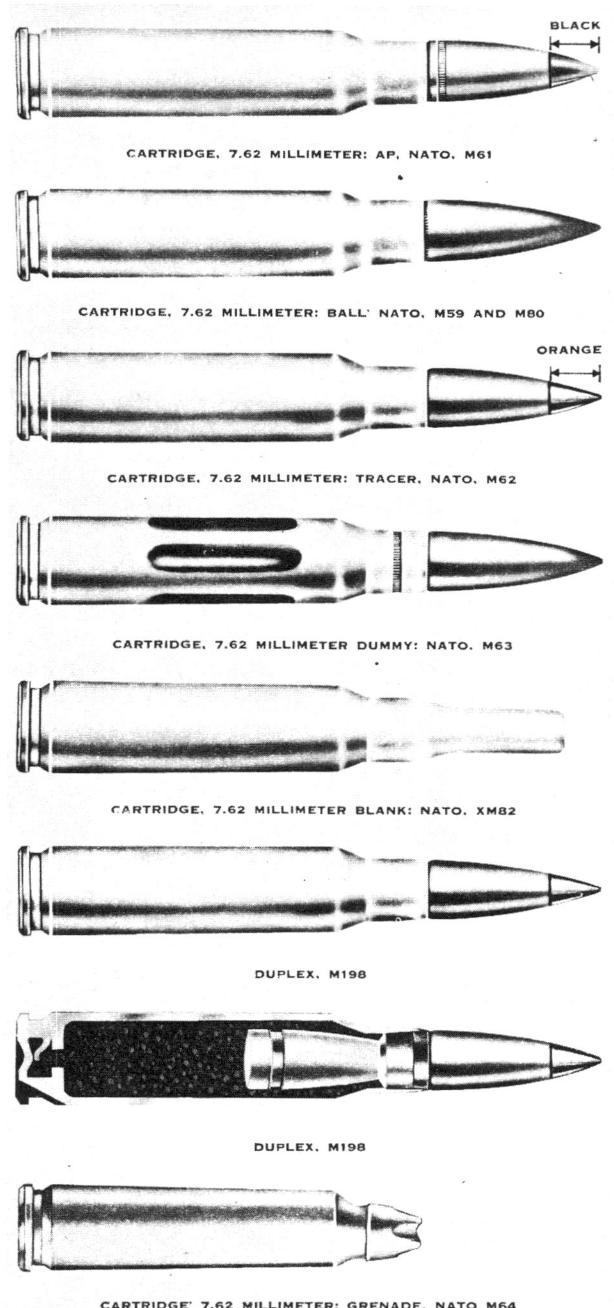

Figure 49. Types of ammunition for the M14 and M14E2 rifles.

CHAPTER 7

ACCESSORIES

33. M2 Bipod

The M2 bipod (fig. 50) is a light, folding mount which clamps onto the gas cylinder and gas cylinder lock of the rifle.

a. Installation (fig. 51). Place the jaws of the yoke assembly so that they encircle the gas cylinder at the gas cylinder lock. Tighten the self-locking bolt with the combination tool, securing the jaws to the gas cylinder.

b. Removal. Using the combination tool, loosen the bolt located beneath the yoke assembly and remove the bipod from the rifle.

***Caution:* Do not remove the cap screw from the jaw assembly.**

34. M6 Bayonet Knife and M8A1 Bayonet Knife Scabbard

The M6 bayonet knife (fig. 52) is utilized for close combat, guarding prisoners and riot control. The M8A1 bayonet scabbard is used to carry the bayonet knife.

a. Installation. Install the bayonet knife to the rifle by alining the groove of the bayonet handle with the bayonet lug on the flash suppressor and the loop of the top portion of the handle on the flash suppressor. Slide the knife rearward until the lugs of the latching lever snap over the bayonet lug (fig. 53).

b. Removal. Grasp the handle of the bayonet and depress the latching lever on the handle, releasing the bayonet lug from the groove in the handle. Slide the bayonet from the rifle.

35. M76 Grenade Launcher

The M76 grenade launcher (fig. 54) is attached to the barrel of the rifle for launching grenades. The barrel of the launcher contains nine annular grooves, numbered 6 to 1, 2A, 3A and 4A. When firing grenades, these are utilized to obtain different ranges by placing the grenade at different positions on the launcher. On the bottom portion of the muzzle end of the launcher, there is a clip-type retainer spring used to hold the grenade on the launcher at the desired position prior to firing. The unmarked groove located above the retainer spring is a safety groove that prevents the grenade from slipping off the launcher if the retainer clip breaks.

a. Installation. To install the grenade launcher, slide the launcher over the flash suppressor. Push the clip latch rearward securing it to the bayonet lug of the flash suppressor (fig. 55).

b. Removal. To remove the grenade launcher, pull downward on the handle of the clip latch, releasing it from the bayonet lug on the flash suppressor, and slide the launcher from the flash suppressor.

36. M15 Grenade Launcher Sight

The grenade launcher sight provides an angular measurement of elevation for firing grenades and can be used for both low angle (direct firing) and high angle firing.

a. Installation. Install the sight to the mounting plate, alining notches of the plate with the click spring tips of the sight (fig. 56). Turn sight clockwise until the index line is alined with the 0° index on the mounting plate. At this position, the leveling bubble should be level. If the bubble cannot be leveled, the rifle should be turned in to the unit armorer.

Note. The mounting plate for the M–15 sight is installed by support maintenance ONLY.

b. Removal. Turn sight counterclockwise until the tips of the clip springs are alined with the notches in the mounting plate; remove the sight from the mounting plate (fig. 56). When not in use, retain the sight in its carrying case.

Note. Removal and mounting of the mounting plate from the stock is accomplished by support maintenance personnel ONLY.

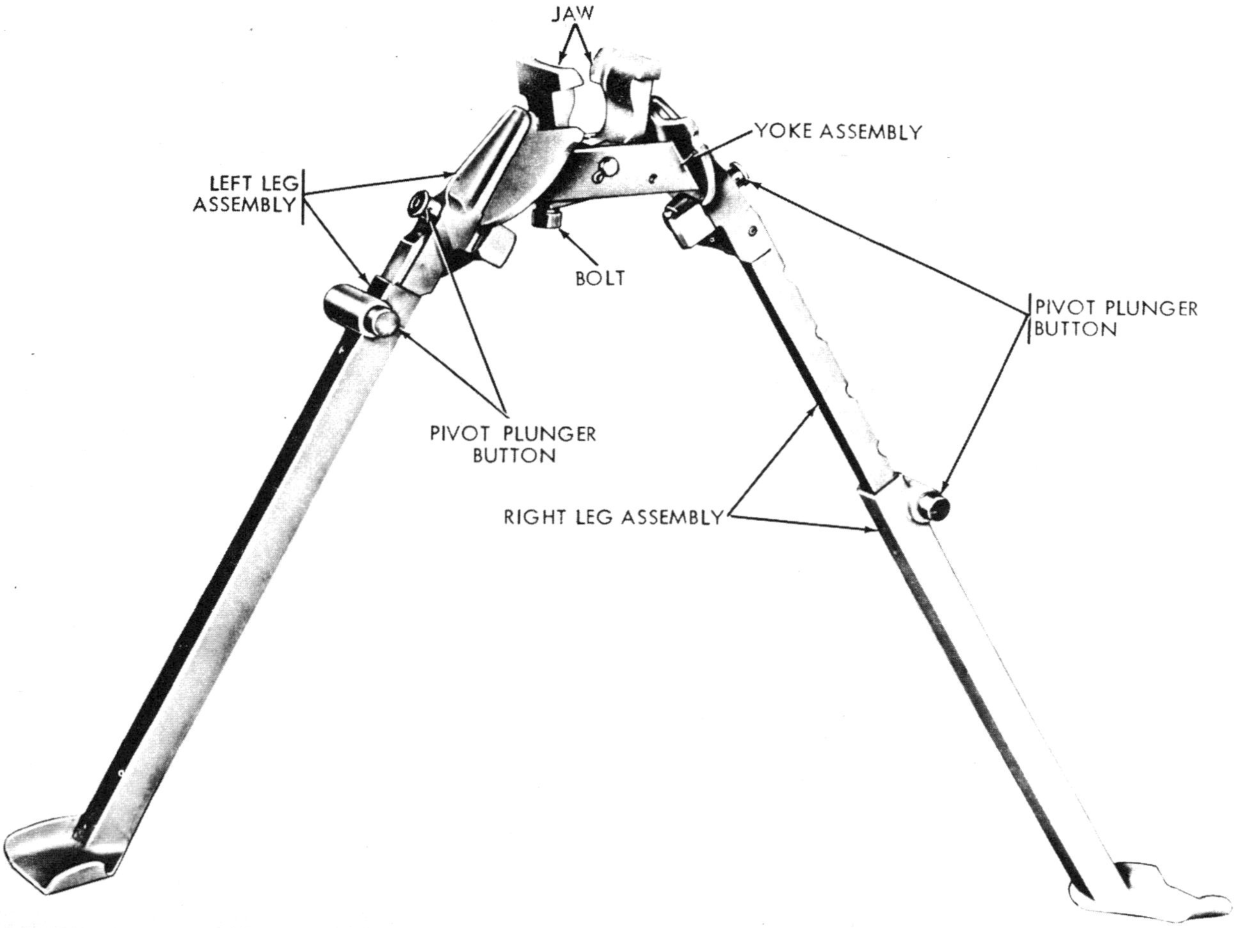

Figure 50. M2 bipod.

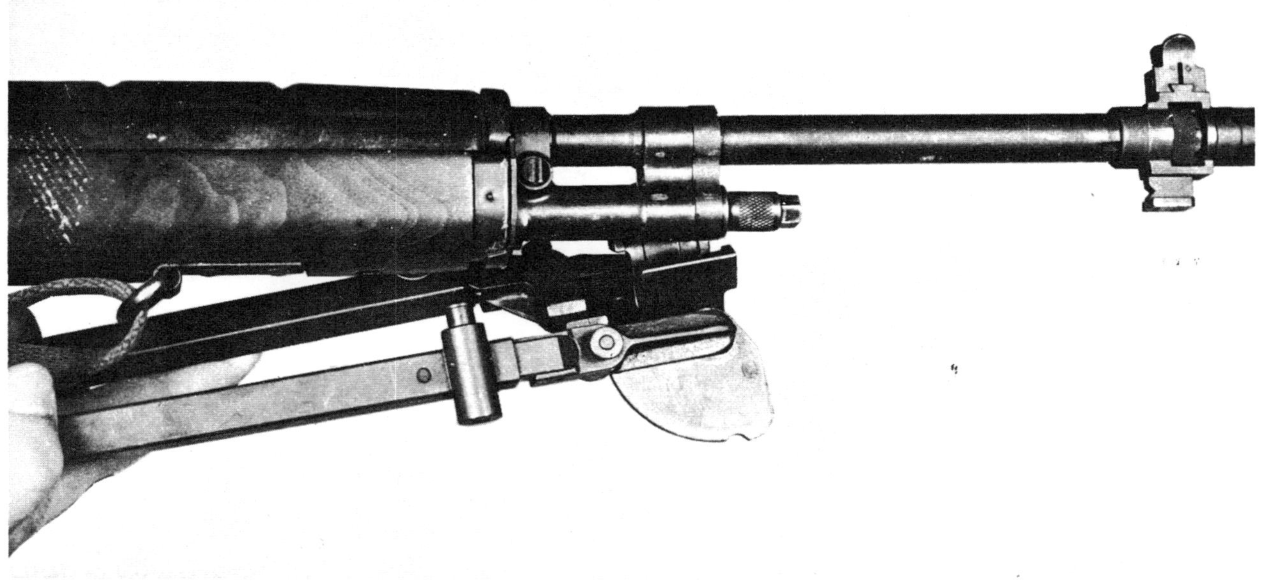

Figure 51. Installation of M2 bipod.

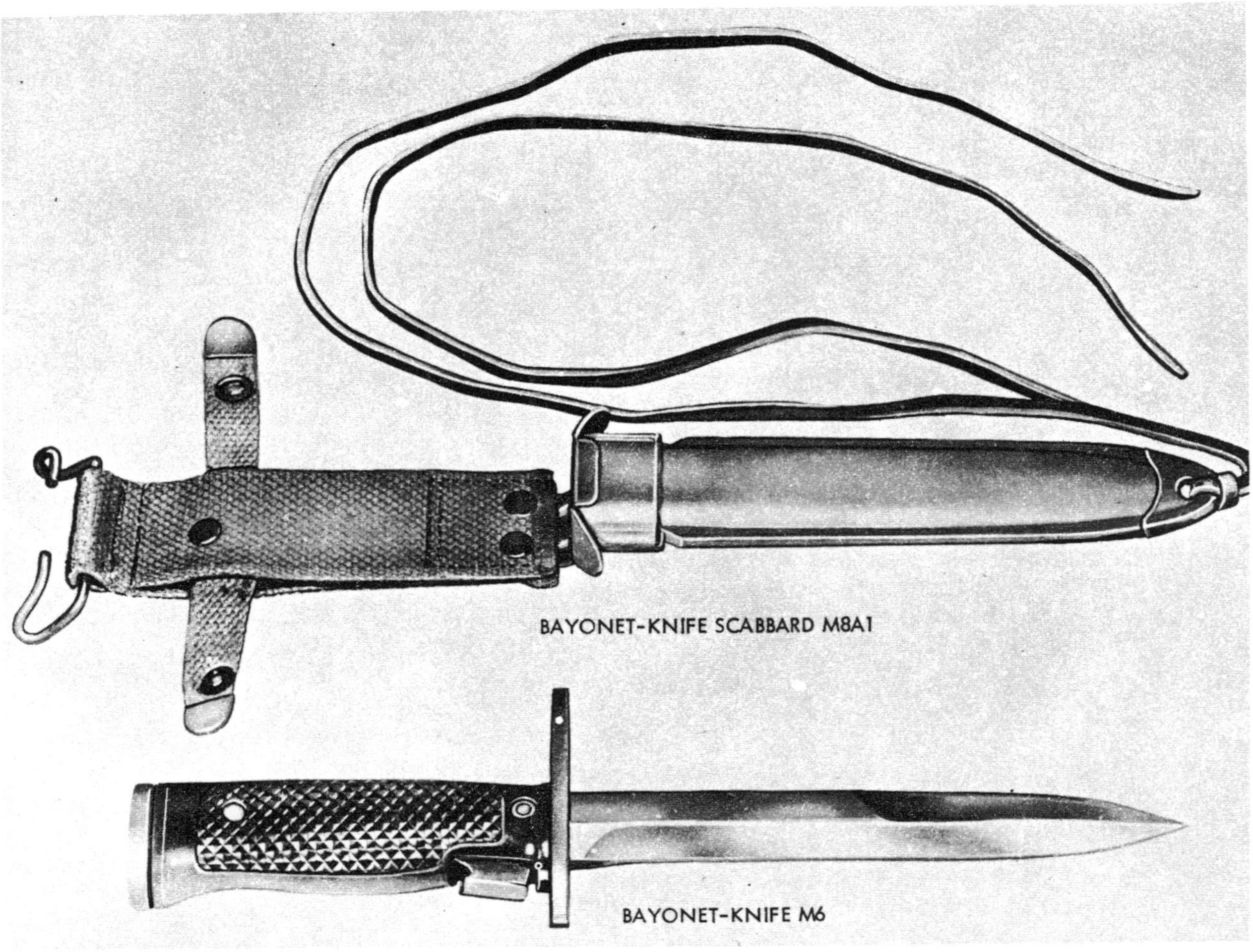

BAYONET-KNIFE SCABBARD M8A1

BAYONET-KNIFE M6

Figure 52. M6 bayonet knife and M8A1 bayonet scabbard.

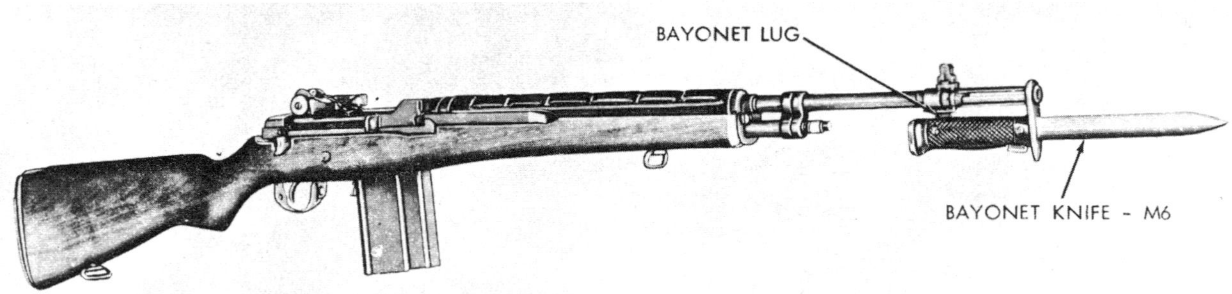

Figure 53. M14 rifle with bayonet knife.

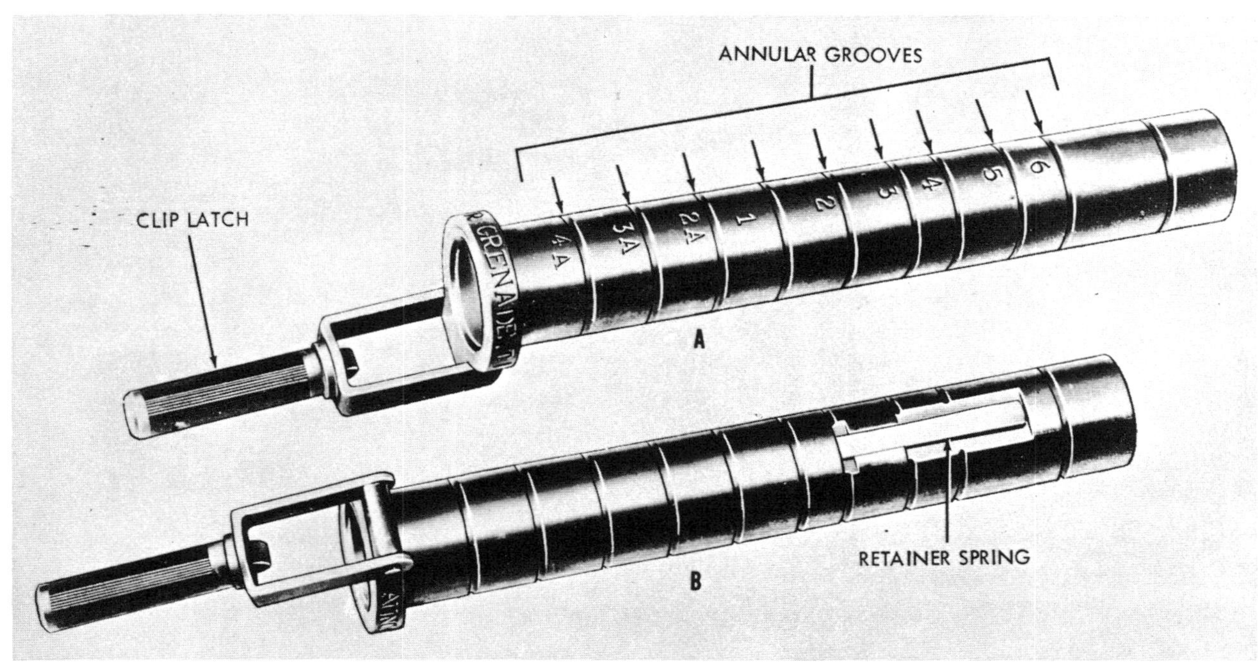

Figure 54. M76 grenade launcher.

Figure 55. M14 rifle with M76 grenade launcher.

37. M12 Blank Firing Attachment and M3 Breech Shield

The blank firing attachment and breech shield (fig. 57) are designed for use when firing blank cartridges. The blank firing attachment, which secures the attachment to the bayonet lug of the flash suppressor, consists of an orifice tube and a spring clip latch. The breech shield, which secures the shield to the cartridge clip guide, is used with the blank firing attachment and consists of a deflector shield and a guide lug with spring plunger.

 a. *Installation* (fig. 58).
  (1) *Blank firing attachment.* Insert the orifice tube in the muzzle opening of the flash suppressor. Pull out on the clip latch and push down on the top of the orifice tube of the blank firing attachment. Release the clip spring latch securing the cut away portion of the latch to the bayonet lug.

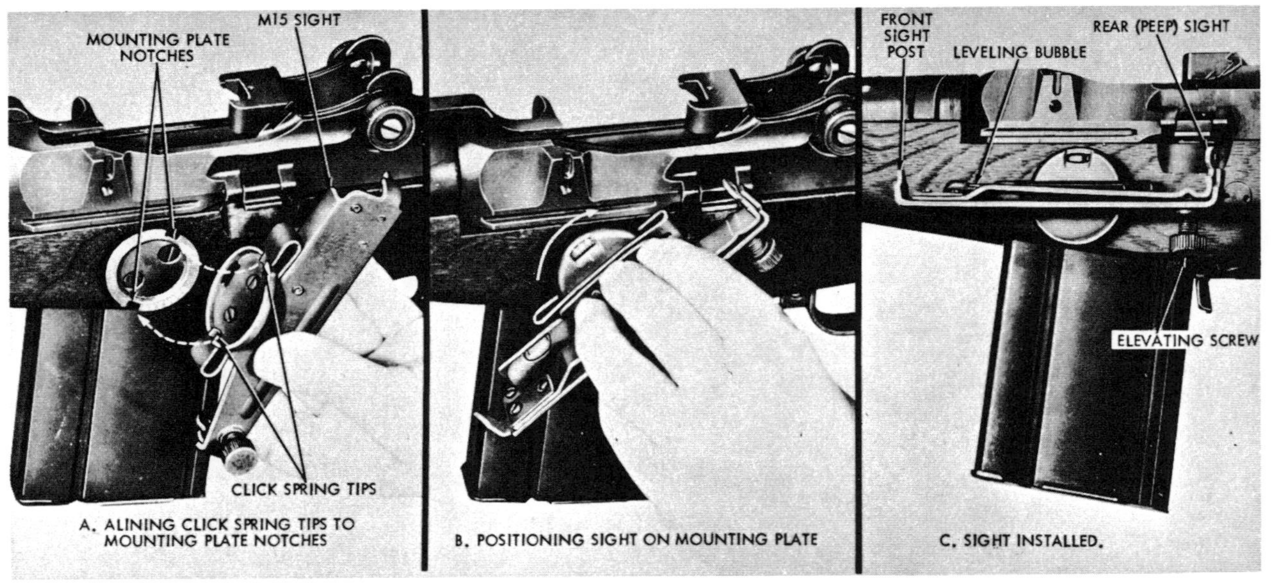

Figure 56. Installation of M15 grenade launcher sight.

(2) *Breech shield.* Insert the guide lug of the breech shield into the slot of the cartridge clip guide. Using an empty blank cartridge, press in on the spring plunger and push down on the breech shield, locking it to the cartridge clip guide.

b. *Removal.*

(1) *Blank firing attachment.* In removing the blank firing attachment from the rifle, pull outward on the spring clip latch releasing it from the bayonet lug. Turn the attachment either to the left or the right of the bayonet lug and slide the attachment from the flash suppressor.

(2) *Breech shield.* Using an empty blank cartridge, or any suitable object, press in on the spring plunger located on the guide lug of the breech shield. Lift the breech shield from the cartridge clip guide.

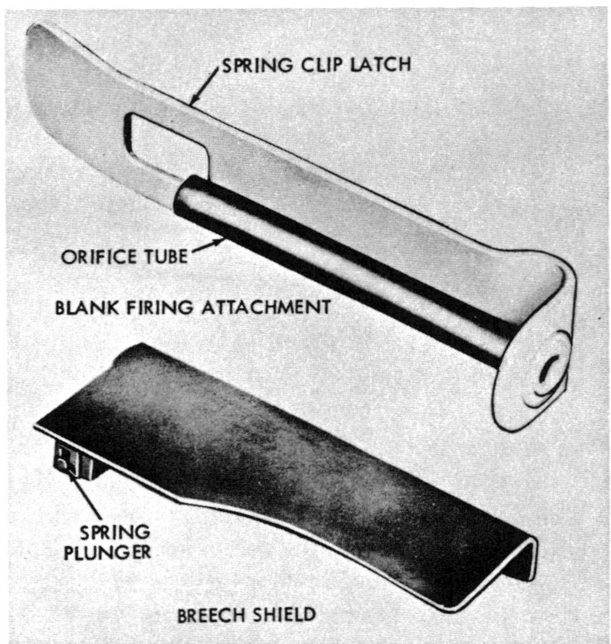

Figure 57. M12 blank firing attachment and M3 breech shield

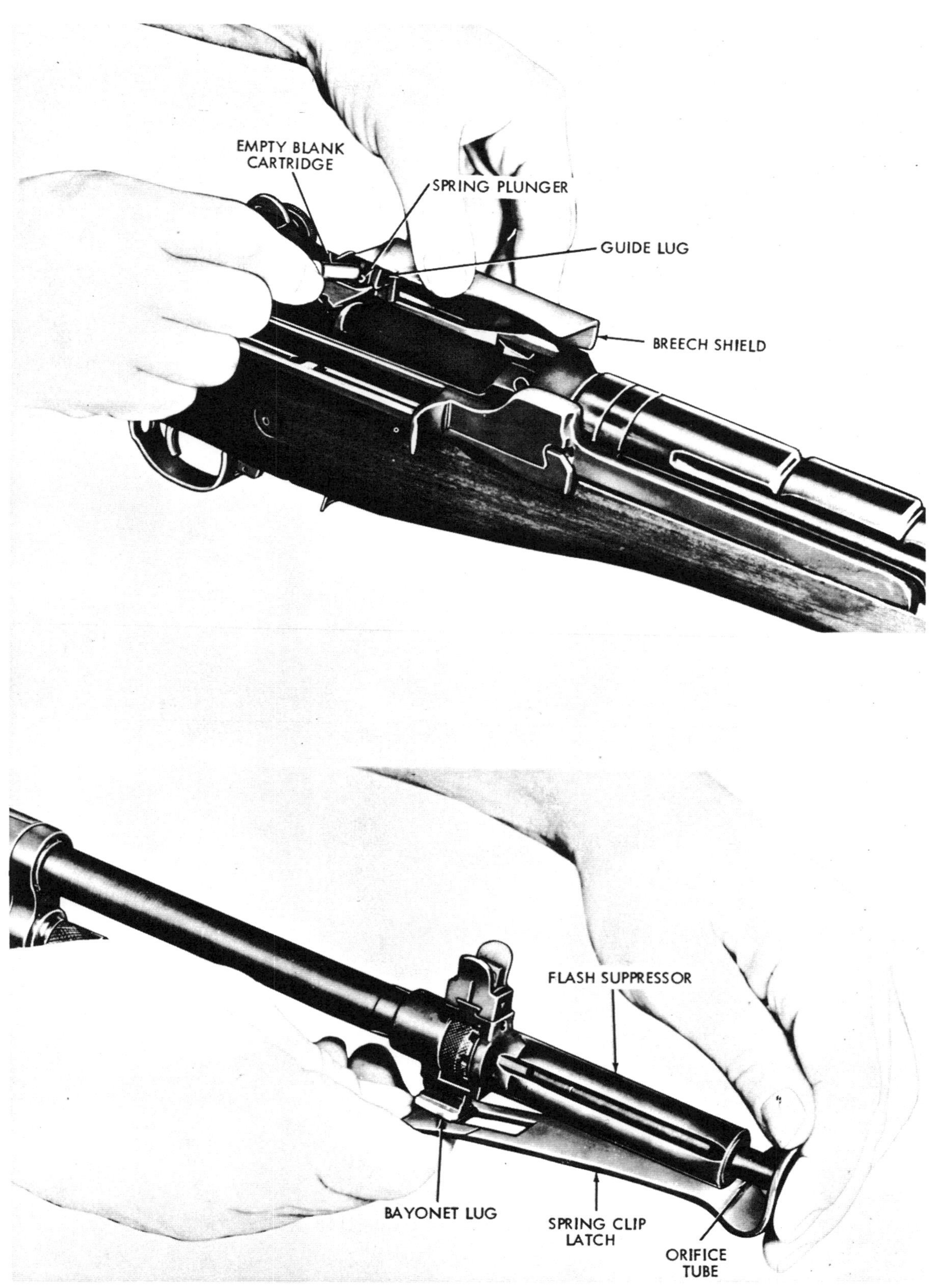

Figure 58. Installation of blank firing attachment and breech shield.

38. Winter Trigger Kit

The winter trigger kit (figs. 59 and 60) is utilized during cold weather and arctic operations by special authorization of the theater commander. It consists of two woodscrews, a winter trigger assembly and a winter safety. The safety can be easily operated by the firer while wearing heavy gloves or mittens because of its long protruding tang which extends approximately one and one-half inches below the firing mechanism.

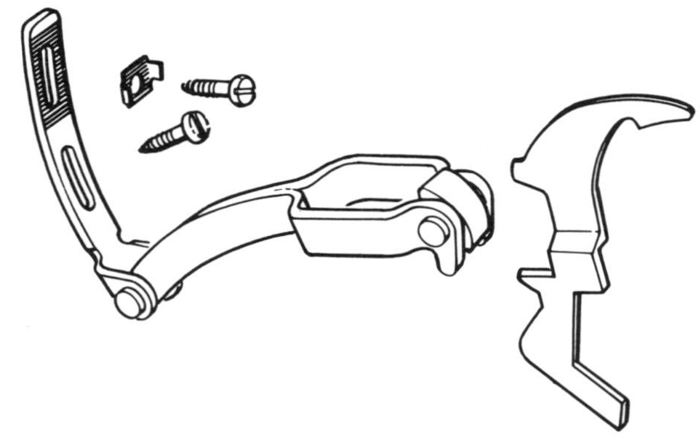

Figure 59. Winter trigger kit.

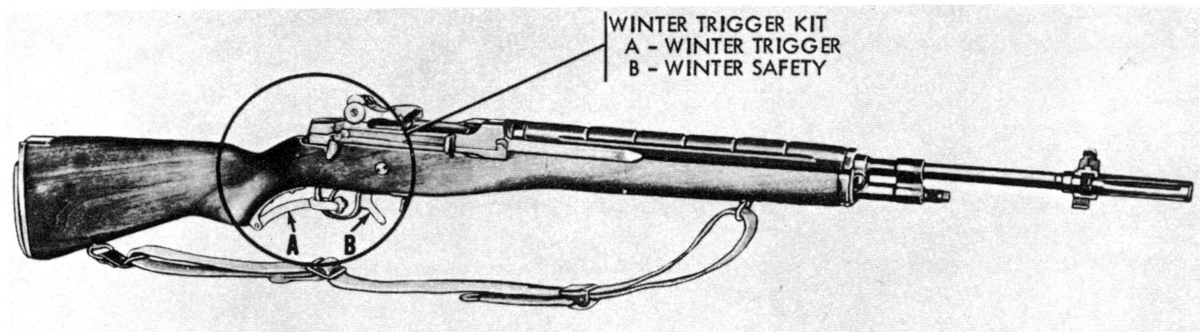

Figure 60. M14 rifle with winter trigger kit.

FM 21–5	Military Training Management.
FM 21–6	Techniques of Military Instruction.
FM 21–40	Small Unit Procedures in Chemical, Biological, and Radiological (CBR) Operations.
FM 22–5	Drill and Ceremonies.
FM 23–71	Rifle Marksmanship.
TM 3–220	Chemical, Biological, and Radiological (CBR) Decontamination.
TM 9–1005–223–12	Operator and Organizational Maintenance Manual 7.62-mm Rifle M14 and Rifle Bipod M2.
TM 9–1005–223–20P	Organizational Maintenance Repair Parts and Special Tool Lists.
TM 9–1305–200	Small-Arms Ammunition.
TM 9–2205	Fundamentals of Small Arms.
TM 38–230	Preservation, Packaging, and Packing of Military Supplies and Equipment.
AR 385–40	Accident Reporting and Records.
AR 385–63	Regulations for Firing Ammunition for Training, Target Practice, and Combat.

NOTE: Chapter 4

Stoppages and Immediate Action, refers to the "unhesitating application of a probable remedy to reduce a stoppage without investigating the cause." This procedure, while warranted in a battlefield situation, is not recommended for civilian use. In the event of a hangfire or misfire, stop firing and engage the safety, keeping the rifle barrel pointed toward your target. Wait fifteen seconds before removing the magazine and unloading the rifle (ejecting any cartridge that might be chambered) and check to see that the barrel is clear. The weapon should be examined, as well as the ammunition. If not absolutely certain of the cause of the failure, it should be examined by a competent gunsmith or returned to the factory.

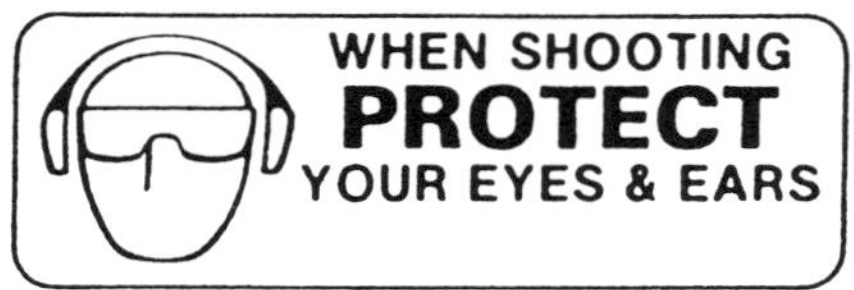

By Order of the Secretary of the Army:

HAROLD K. JOHNSON,
General, United States Army,
Chief of Staff.

Official:
J. C. LAMBERT,
Major General, United States Army,
The Adjutant General.

Distribution:
Active Army:

DCSPER (2)	LOGCOMD (1)	MFSS (1)
DCSPR (2)	USACDC (2)	USAOC&S (1)
ACSI (2)	Armies (25)	USAQMS (1)
DSCLOG (2)	Corps (3)	USASCS (1)
DCSOPS (2)	Div (10	USACHS (1)
CORC (2)	Div Arty (5)	USAES (1)
CRD (1)	Bde (5)	USATSCH (1)
COA (1)	Regt (5)	USACMLCS (1)
CINFO (1)	Gp (1)	USASESCS (1)
TIG (1)	BG (5)	USMA (2)
TJAGSA (1)	CC (5)	Svc Colleges (5)
CNGB (2)	Bn (5)	Mil Msn (1)
ACSFOR (2)	Co/Btry (5)	USATC (10) except
USCONARC (5)	Br Svc Sch (5) except	USATC Inf (25)
ARADCOM (2)	USAMPS (1)	
ARADCOM Rgn (1)		

NG: State AG (3) ; units—same as Active Army except allowance is four copies to each unit.
USAR: Units—same as Active Army except allowance is two copies to each unit.
For explanation of abbreviations used, see AR 320–50.

NOTES

Other Books Available From Desert Publications

No.	Title	Price
001	Firearms Silencers Volume 1	$9.95
003	The Silencer Cookbook	$9.95
004	Select Fire Uzi Modification Manual	$9.95
005	Expedient Hand Grenades	$16.95
007	007 Travel Kit, The	$8.00
008	Law Enforcement Guide to Firearms Silencer	$8.95
009	Springfield Rifle, The	$11.95
010	Full Auto Vol 3 MAC-10 Mod Manual	$9.95
012	Fighting Garand, The	$11.95
013	M1 Carbine Owners Manual	$9.95
014	Ruger Carbine Cookbook	$11.95
015	M-14 Rifle, The	$9.95
016	AR-15, M16 and M16A1 5.56mm Rifles	$11.95
017	Shotguns	$11.95
019	AR15 A2/M16A2 Assault Rifle	$8.95
022	Full Auto Vol 7 Bingham AK-22	$9.95
027	Full Auto Vol 4 Thompson SMG	$9.95
030	STANAG Mil-Talk	$12.95
031	Thompson Submachine Guns	$13.95
033	H&R Reising Submachine Gun Manual	$12.95
035	How to Build Silencers	$6.95
036	Full Auto Vol 2 Uzi Mod Manual	$9.95
049	Firearm Silencers Vol 3	$13.95
050	Firearm Silencers Vol 2	$19.95
054	Company Officers HB of Ger. Army	$11.95
056	German Infantry Weapons Vol 1	$14.95
058	Survival Armory	$27.95
060	Survival Gunsmithing	$9.95
061	FullAuto Vol 1 Ar-15 Mod Manual	$8.95
064	HK Assault Rifle Systems	$27.95
065	SKS Type of Carbines, The	$16.95
066	Private Weaponeer, The	$9.95
067	Rough Riders, The	$24.95
068	Lasers & Night Vision Devices	$29.95
069	Ruger P-85 Family of Handguns	$14.95
071	Dirty Fighting	$11.95
072	Live to Spend It	$29.95
073	Military Ground Rappelling Techniques	$11.95
074	Smith & Wesson Autos	$27.95
080	German MG-34 Machinegun Manual	$9.95
081	Crossbows/ From 35 Years With the Weapon	$11.95
082	Op. Man. 7.62mm M24 Sniper Weapon	$7.95
083	USMC AR-15/M-16 A2 Manual	$16.95
084	Urban Combat	$21.95
085	Caching Techniques of U.S. Army Spec Forces	$9.95
086	US Marine Corps Essential Subjects	$16.95
087	The L'il M-1, The .30 Cal. M-1 Carbine	$14.95
088	Concealed Carry Made Easy	$14.95
089	Apocalypse Tomorrow	$17.95
090	M14 and M14A1 Rifles & Rifle Marksmanship	$16.95
091	Crossbow As a Modern Weapon	$11.95
092	MP40 Machinegun	$11.95
093	Map Reading and Land Navigation	$19.95
094	U. S. Marines Scout/Sniper Training Manual	$16.95
095	Clear Your Record & Own a Gun	$14.95
096	Sig Handguns	$16.95
097	Poor Man's Nuclear Bomb	$19.95
098	Poor Man's Sniper Rifle	$14.95
100	Submachine Gun Designers Handbook	$16.95
101	Lock Picking Simplified	$8.50
102	Combination Lock Principles	$8.95
103	How to Fit Keys by Impressioning	$7.95
104	Keys To Understanding Tubular Locks	$9.95
105	Techniques of Safe & Vault Munipulation	$9.95
106	Lockout -Techniques of Forced Entr	$11.95
107	Bugs Electronic Surveillance	$11.95
110	Improvised Weapons of Amer. Undergrnd	$10.00
111	Training Hndbk of the American Underground	$10.00
114	FullAuto Vol 8 M14A1 & Mini 14	$9.95
116	Handbook Bomb Threat/Search Procedures	$8.00
117	Improvised Lock Picks	$10.00
119	Fitting Keys By Reading Locks	$7.00
120	How to Open Handcuffs Without Keys	$9.95
121	Electronic Locks Volume 1	$8.00
122	With British Snipers, To the Reich	$24.95
125	Browning Hi-Power Pistols	$9.95
126	P-08 Parabellum Luger Auto Pistol	$9.95
127	Walther P-38 Pistol Manual	$9.95
128	Colt .45 Auto Pistol	$9.95
129	Beretta - 9MM M9	$11.95
130	FullAuto Vol 5 M1 Carbine to M2	$9.95
132	Ranger Handbook	$16.95
133	FN-FAL Auto Rifles	$13.95
135	AK-47 Assault Rifle	$11.95
136	UZI Submachine Gun	$9.95
139	USMC Battle Skills Training Manual	$49.95
140	Sten Submachine Gun, The	$9.95
141	Terrorist Explosives Handbook	$6.95
142	U. S. Army Counterterrorism Training Manual	$14.95
143	Sniper Training	$24.95
144	The Butane Lighter Handgrenade	$9.95
146	The Official Makarov Pistol Manual	$12.95
147	Official Makarov 9mm Pistol Manual	$12.95
148	Unarmed Against the Knife	$9.95
149	Black book of Booby Traps	$14.95
151	Militia Battle Manual	$18.95
152	Glock's Handguns	$17.95
153	Heckler and Koch's Handguns	$17.95
154	The Poor Man's R. P. G.	$17.95
155	The Poor Man's Ray Gun	$9.95
156	The Anachist Handbook Vol. 2	$11.95
157	How to Make Disposable Silencers Vol. 2	$16.95
158	Modern Day Ninjutsu	$14.95
159	The Squeaky Wheel	$12.95
160	The Sicilian Blade	$13.95
162	The Anarchist Handbook Vol. 1	$11.95
163	How to Make Disposable Silencers Vol. 1	$16.95
164	The Anarchist Handbook Vol. 3	$11.95
165	How to Build Practical Firearm Silencers	$14.95
166	Full Auto Modification Manual	$18.95
167	Black Bokk of Revenge	$13.95
168	How to Train a Guard Dog	$13.95
172	Bankruptcy, Credit and You	$14.95
173	Mercenary's Tactical Handbook	$14.95
174	Killer Tactics	$14.95
177	Save More So You Can Spend More	$14.95
178	Surviving Global Slavery	$13.95
180	Build Your Own AR-15	$14.95
181	Credit Improvement & Protection Hndbk	$19.95
183	Black Book of Arson	$19.95
184	Entrapment	$19.95
185	How to Live Safely in a Dangerous World	$17.95
186	The FN-FAL et al	$18.95
187	Basic Locksmithing	$19.95
188	The Mental Edge - revised	$17.95
189	The Survival Bible	$99.95
190	The Criminal Defendants Bible	$49.95
191	Internet Predator	$17.95
200	Fighting Back on the Job	$11.95
202	Secret Codes & Ciphers	$9.95
204	Improvised Munitions Black Book Vol 1	$14.95
205	Improvised Munitions Black Book Vol 2	$14.95
206	CIA Field Exp Preparation of Black Powder	$8.95
207	CIA Field Exp. Meth/Explo. Preparat	$8.95
209	CIA Improvised Sabotage Devices	$12.00
210	CIA Field Exp. Incendiary Manual	$12.00
211	Science of Revolutionary Warfare	$9.95
212	Agents HB of Black Bag Ops.	$9.95
214	Electronic Harassment	$11.95
217	Improvised Rocket Motors	$6.95
218	Impro. Munitions/Ammonium Nitrate	$9.95
219	Improvised Batteries/Det. Devices	$8.95
220	Impro. Explo/Use In Deton. Devices	$7.95
221	Evaluation of Imp Shaped Charges	$8.95
222	American Tools of Intrigue	$12.00
225	Impro. Munitions Black Book Vol 3	$23.95
226	Poor Man's James Bond Vol 2	$24.95
227	Explosives and Propellants	$11.95
229	Select Fire 10/22	$11.95
230	Poor Man's James Bond Vol 1	$24.95
231	Assorted Nasties	$19.95
232	Full Auto Modification Manual	$18.95
234	L.A.W. Rocket System	$8.00
240	Clandestine Ops Man/Central America	$11.95
241	Mercenary Operations Manual	$9.95
250	Improvised Shaped Charges	$8.95
251	Two Component High Exp. Mixtures	$9.95
260	Survival Evasion & Escape	$13.95
262	Infantry Scouting, Patrol. & Sniping	$13.95
263	Engineer Explosives of WWI	$7.95
300	Brown's Alcohol Motor Fuel Cookbook	$13.95
301	How to Build a Junkyard Still	$11.95
303	Alcohol Distillers Handbook	$17.95
306	Brown's Book of Carburetors	$11.95
310	MAC-10 Cookbook	$9.95
350	Cheating At Cards	$12.00
367	Brown's Lawsuit Cookbook	$18.95
400	Hand to Hand Combat	$11.95
401	USMC Hand to Hand Combat	$7.95
402	US Marine Bayonet Training	$8.95
404	Camouflage	$12.95
409	Guide to Germ Warfare	$13.95
410	Emergency War Surgery	$24.95
411	Homeopathic First Aid	$9.95
412	Defensive Shotgun	$14.95
414	Hand to Hand Combat by D'Eliscue	$7.95
415	999 Survived	$6.00
416	Sun, Sand & Survival	$6.00
420	USMC Sniping	$16.95
424	Prisons Bloody Iron	$13.95
425	Napoleon's Maxims of War	$9.95
429	Invisible Weapons/Modern Ninja	$11.95
432	Cold Weather Survival	$11.95
435	Homestead Carpentry	$10.00
436	Construction Secret Hiding Places	$11.95
437	US Army Survival	$29.95
438	Survival Shooting for Women	$11.95
440	Survival Medicine	$11.95
442	Can You Survive	$12.95
443	Canteen Cup Cookery	$9.95
444	Leadership Hanbook of Small Unit Ops	$11.95
445	Vigilante Handbook	$11.95
447	Shootout II	$14.95
448	Catalog of Military Suppliers	$14.95
454	Survival Childbirth	$8.95
455	Police Karate	$10.95
456	Survival Guns	$21.95
457	Water Survival Training	$5.95
470	Emergency Medical Care/Disaster	$11.95
500	Guerilla Warfare	$14.95
504	Ranger Training & Operations	$14.95
507	Spec. Forces Demolitions Trng HB	$16.95
508	IRA Handbook	$9.95
510	Battlefield Analysis/Inf. Weapons	$9.95
511	US Army Bayonet Training	$7.95
512	Desert Storm Weap. Recog. Guide	$9.95
542	Professional Homemade Cher Bomb	$10.95
544	Combat Loads for Sniper Rifles	$12.00
551	Take My Gun..If You Dare	$10.95
552	Aunt Bessie's Wood Stove Cookbook	$7.50
605	Jackedup & Ripped off	$10.00
610	Trapping & Destruc. of Exec. Cars	$10.00
C-002	How To Open A Swiss Bank Account	$6.95
C-011	Defending Your Retreat	$9.95
C-019	How To Become A Class 3 MG Dealer	$3.95
C-020	Methods of Long Term Storage	$8.95
C-021	How To Obtain Gun Dealer Licenses	$3.95
C-022	Confid. Gun Dealers Guide/Whlesler	$2.50
C-023	Federal Firearms Laws	$4.50
C-028	Militarizing the Mini-14	$7.95
C-029	M1 Carbine Arsenal History	$6.95
C-030	OSS/CIA Assassination Device	$4.00
C-038	Hw To Build A Beer Can Morter	$4.95
C-040	Criminal Use of False ID	$11.95
C-050	Surviving Doomsday	$11.95
C-052	CIA Explosives for Sabotage	$9.00
C-058	Brass Knuckle Bible	$7.95
C-080	USA Urban Survival Arsenal	$7.95
C-085	Dead or Alive	$7.95
C-099	Elementary Field Interrogation	$9.95
C-175	Beat the Box	$7.95
C-177	Boobytraps	$8.95
C-209	Guide/Vietcong Boobytraps/Device	$8.00
C-386	Self-Defense Requires No Apology	$11.95
C-477	A Rifle man Went To War	$24.95
C-679	M16A1 Rifle Manual Cartoon Version	$6.95
C-9281	How to Build Your Own AR-15	$14.95

PRICES SUBJECT TO CHANGE WITHOUT NOTICE

870-862-2077

Desert Publications
215 S. Washington, Dept. BK-015
El Dorado, AR 71730 USA

info@deltapress.com

Shipping & handling
1 item $4.50
2 or more $6.50